THE AI-READY CLASSROOM

Preparing Schools and Educators to Teach Artificial Intelligence Responsibly

Susie Hala

AI Educator | Author | Founder of the IntelliGloss AI Education Series

IntelliGloss Press

Disclaimer

Educational Purpose Disclaimer

"This publication is part of the IntelliGloss AI Education Series, a structured curriculum support system for AI literacy in grades 6–12."

The information presented in *The AI-Ready Classroom: Preparing Schools and Educators to Teach Artificial Intelligence Responsibly* is provided for **educational and informational purposes only**. This book is intended to support educators, administrators, parents, and learners in understanding the concepts, opportunities, and challenges associated with artificial intelligence in education. It is not intended to serve as legal, technical, policy, cybersecurity, or professional advice.

Readers should consult qualified professionals, institutional guidelines, and local regulations before implementing any educational programs, technologies, or policies described in this book.

Technology and Accuracy Disclaimer

Artificial intelligence is a rapidly evolving field. While every effort has been made to ensure the accuracy and reliability of the information presented at the time of publication, technologies, policies, research findings, and best practices may change over time. The author and publisher make **no guarantees that all information will remain current or applicable in future technological environments.**

Readers are encouraged to verify tools, platforms, and recommendations independently before using them in educational settings.

Implementation Responsibility

Schools, educators, and institutions are solely responsible for determining how and whether to implement artificial intelligence tools, strategies, or classroom activities discussed in this book. Educational environments vary widely, and policies regarding technology use, student privacy, and curriculum development differ by district, state, and country.

The author and publisher assume **no responsibility for outcomes resulting from the implementation of ideas, examples, or frameworks described in this book.**

Student Safety and Data Privacy

Artificial intelligence tools may involve the collection or processing of data. Educators and institutions are responsible for ensuring that any technology used complies with applicable

student privacy laws, data protection regulations, and school policies, including but not limited to FERPA, COPPA, GDPR, or other relevant regulations depending on jurisdiction.

The author and publisher are **not responsible for the privacy practices, security policies, or data handling procedures of third-party platforms or software mentioned in this book.**

Third-Party Tools and References

This book may reference third-party technologies, organizations, research studies, or platforms for illustrative and educational purposes. Such references do not constitute endorsements or guarantees regarding the performance, reliability, safety, or suitability of those tools for educational use.

Educators and institutions should conduct their own evaluations and risk assessments before adopting any technology.

No Professional Liability

The author and publisher shall not be held liable for any direct, indirect, incidental, or consequential damages resulting from the use or misuse of the information contained in this book. This includes but is not limited to decisions related to curriculum design, technology adoption, classroom instruction, institutional policies, or educational outcomes.

Educational Interpretation

The concepts and examples provided in this book are intended to support discussion and learning. They should not be interpreted as official standards, mandatory guidelines, or universally applicable policies for artificial intelligence education.

Schools and educators are encouraged to adapt the ideas presented here according to their **local policies, cultural context, educational standards, and student needs.**

Author's Intent

The goal of this book is to promote **responsible AI literacy, thoughtful discussion, and ethical awareness** as schools prepare students for a world increasingly shaped by intelligent technologies.

The author encourages readers to approach artificial intelligence education with curiosity, caution, and a commitment to responsible learning.

Title: *The AI-Ready Classroom: Preparing Schools and Educators to Teach Artificial Intelligence Responsibly*

ISBN: 978-1-972925-01-0

Published by
IntelliGloss Press United States

This book is intended for educational and informational purposes. Artificial intelligence technologies and policies evolve rapidly, and the author makes no guarantee regarding the completeness, accuracy, or continued applicability of the information presented. Educators, schools, and institutions are responsible for evaluating and implementing technologies in accordance with their own policies, legal requirements, and educational standards. Cover design and layout by IntelliGloss Press

First Edition 2026

Printed in the United States of America

Author

Susie Hala
AI Educator and Author

Dedication

To the teachers who inspire curiosity, courage, and lifelong learning.

Your classrooms are where the future begins. As technology transforms our world, it is educators who help students ask the right questions, think critically, and use knowledge with wisdom and responsibility.

This book is also dedicated to the students who will grow up in a world shaped by artificial intelligence. May you learn not only how technology works, but also how to guide it with ethics, creativity, and humanity.

And with love to my granddaughter, **Raquel**, whose curiosity about the world reminds me every day why education matters. May your generation learn to build a future where intelligence—both human and artificial—serves the good of all.

Preface

Two Students. Two Paths. One Future.

Before we begin, it is important to understand why learning how artificial intelligence works—not just how to use it—matters for every student.

Imagine two students who have just graduated from high school and are preparing for a job interview. The first student knows how to use artificial intelligence tools. This student can generate answers, complete assignments quickly, and rely on AI to assist with tasks. However, this student does not understand how artificial intelligence works behind the scenes.

The second student also knows how to use AI tools, but in addition, understands the foundational concepts behind them. This student has learned about data, algorithms, tokens, models, and how intelligent systems are trained and operate. This student can think critically about AI, recognize its limitations, and adapt when needed.

At first, the first student may appear faster and more efficient. However, over time, the second student demonstrates a deeper level of understanding, independence, and problem-solving ability. The second student is not only able to use artificial intelligence but also to question it, improve it, and grow with it.

This book is written for the second student.

In a world where artificial intelligence is becoming part of everyday life, students must not only learn how to use intelligent systems—they must learn how they work. True understanding leads to confidence, responsibility, and the ability to lead in an AI-driven future.

This perspective is not just theoretical—it is personal.

There was a time when I was the first student.

I knew how to use systems. I followed processes, completed tasks, and worked within structured environments that required accuracy and consistency. At the time, I did not think of these systems as artificial intelligence. They were simply part of the job—tools that had to be learned, followed, and mastered.

But looking back, I now understand something much deeper. I was working within rule-based systems—the earliest foundation of what we now recognize as artificial intelligence.

That realization changed everything.

As I began studying modern AI, including machine learning, deep learning, and intelligent systems, I started to see a clear connection between the past and the present. The structured systems I once worked in, the logic behind decision-making, and the step-by-step processes all still exist today. They have simply evolved.

Artificial intelligence did not appear overnight. It grew from rule-based systems into systems that can now learn, adapt, and assist in ways we never imagined.

I made a decision to move from simply using systems to understanding them.

That journey—from student one to student two—is the reason these books exist.

My mission is to help others bridge that same gap.

I believe that artificial intelligence should be understandable for everyone. It should not feel intimidating or out of reach. Students should feel confident exploring it. Teachers should feel prepared to introduce it. Everyday learners should have access to clear, structured explanations that make sense.

You do not need a background in technology. You do not need to know how to code. You only need the right explanation.

The IntelliGloss Approach

This book is part of the **IntelliGloss AI Education Series**, which is designed to simplify complex ideas and make them accessible to a wide range of learners.

Each book follows a structured approach:

clear explanations real-world

connections step-by-step

learning practical

applications

Because true understanding does not come from complexity—it comes from clarity.

A Message to Students and Readers

If you have ever felt that technology was too complicated…
If you have ever been confused by systems that were never clearly explained…

I want you to know:

You are capable of understanding artificial intelligence.

Sometimes, the challenge is not the concept itself—it is how it has been presented.

This series was created to change that.

A Message to Educators

To the educators using this material:

These books are designed to support you in introducing artificial intelligence in a way that is clear, structured, and meaningful for students.

AI is becoming part of every industry, and students deserve more than just exposure to tools—they deserve understanding.

This series aims to provide that foundation.

Final Reflection

Every experience we have—especially the challenging ones—prepares us for something greater.

The systems I once worked in taught me discipline, structure, and attention to detail. Today, those same lessons allow me to break down complex ideas and make them understandable for others.

That is what this series represents:

Turning complexity into clarity.

About the IntelliGloss AI Education Series
A Complete Learning System for the AI-Ready Classroom

The IntelliGloss AI Education Series is a comprehensive collection of educational books designed to help students, educators, and school districts understand artificial intelligence in a clear, structured, and responsible way. As artificial intelligence continues to reshape industries, education, and daily life, there is an urgent need for learning materials that go beyond simply using AI tools and instead explain how AI actually works.

Today, many students are already using artificial intelligence in their daily lives. They interact with chatbots, generate content, search for answers, and complete assignments with the help of AI-powered tools. However, while students are becoming increasingly skilled at using AI, most do not understand what is happening behind the scenes. The same is true for many educators, who are being asked to guide students through a rapidly evolving technological landscape without access to clear, structured resources that explain the foundations of AI.

This gap between using AI and understanding AI represents one of the most important challenges in modern education.

The IntelliGloss AI Education Series was created to close that gap.

Each book within the IntelliGloss AI Education Series is intentionally structured to align with real classroom instruction and academic pacing. Every title is designed to support a full semester of learning, with sixteen chapters that correspond to a typical sixteen-week school schedule. This structure allows educators to teach artificial intelligence in a consistent, organized, and manageable way without the need to redesign their curriculum.

To ensure clarity, consistency, and depth of understanding, every book in the series is built on three core sections:

The first section is the **Family Section**, located at the front of each book. This section introduces the foundational "families" of artificial intelligence, beginning with the concept of the **Ingredients of Artificial Intelligence**—a simple and approachable way to understand how AI systems are built. From this foundation, students are introduced to key concepts such as data, algorithms, tokens, parameters, memory, and infrastructure. These families serve as the building blocks of AI, providing a clear mental framework before deeper learning begins.

The second section is the **Body of the Book**, consisting of sixteen chapters designed for step-bystep learning throughout the semester. Each chapter builds upon the previous one and includes structured educational components such as summaries, reflections, worksheets, quizzes, and answer keys. This ensures that students are not only exposed to concepts but are also able to understand, apply, and retain what they learn.

The third section is the **AI Glossary**, located at the back of each book. This glossary reinforces learning by providing clear definitions of key terms introduced throughout the text. It also connects directly to the series' foundational reference work, allowing students and educators to deepen their understanding beyond the classroom lessons.

At the core of this entire series is a foundational reference work:

IntelliGloss: The Definitive AI Dictionary

This foundational dictionary is available in print format, with expanded distribution across national and international platforms.

This flagship dictionary serves as the backbone of all IntelliGloss publications. With over 1,100 pages of clearly explained AI terms, concepts, and systems, it provides a unified language for understanding artificial intelligence—from basic building blocks such as bits and algorithms to advanced systems such as neural networks, large language models, and global AI infrastructure.

Every book in this series builds upon the definitions, frameworks, and structured thinking established in this dictionary, ensuring consistency, clarity, and depth across all learning materials.

The IntelliGloss AI Education Series is organized as a progressive learning system, guiding students from foundational knowledge to more advanced concepts. These books are not standalone resources; they are part of a complete educational pathway designed to build true AI literacy—not just the ability to use AI tools, but the understanding of how artificial intelligence systems function, make decisions, and impact the world.

At the time of this publication, multiple titles within the series have been completed, with many more currently in development. Together, they form a growing library of AI curriculum materials intended to support schools nationwide.

This is not just a collection of books—it is a complete system designed to help schools prepare students for a future where understanding artificial intelligence is no longer optional, but essential.

Who This Book Is For

This book is written for the professionals responsible for guiding education in an age of rapidly advancing technology.

The primary audience includes:

School districts and unified school districts seeking a structured approach to artificial intelligence literacy.

Teachers and classroom educators working with students in grades 6–12 who are encountering AI tools both inside and outside the classroom.

School administrators and principals responsible for developing policies that balance innovation, academic integrity, and student safety.

Curriculum developers and instructional leaders who are designing learning programs that prepare students for an increasingly technology-driven world.

Artificial intelligence is already influencing how students research, write, solve problems, and communicate. Yet many educators have not received formal training in how these systems function or how they should be addressed in the classroom.

This book is designed to bridge that gap.

It provides educators with clear explanations, practical frameworks, and implementation strategies that support responsible and thoughtful integration of artificial intelligence into education.

The goal is not to replace existing curriculum. The goal is to help schools strengthen critical thinking, digital literacy, and ethical awareness as intelligent technologies become more present in students' academic and professional futures.

Whether a school district is beginning to explore AI policies or already experimenting with AIsupported learning tools, this guide provides a structured path forward.

Artificial intelligence will continue evolving. With the right knowledge and leadership, educators can ensure that students learn not only how to use these technologies, but how to understand them, question them, and apply them responsibly.

How School Districts Can Use This Book

Artificial intelligence is developing faster than most traditional curriculum cycles. As a result, many school districts are seeking practical ways to address AI in a responsible and structured manner without disrupting existing academic programs.

This book is designed to serve as both an **educational resource and an implementation guide** for school districts exploring artificial intelligence literacy.

Districts may use this book in several ways.

1. Professional Development for Educators

School districts can use this guide as a professional learning resource for teachers and instructional staff. The chapters provide foundational explanations of artificial intelligence, classroom considerations, and instructional strategies that help educators understand how AI systems work and how students are already encountering them.

Professional development sessions may include:

• Faculty workshops introducing AI literacy concepts

- Teacher discussion groups exploring classroom scenarios
- Training sessions focused on responsible AI use in education
- Collaborative planning meetings to adapt assignments and assessments

By building educator understanding first, schools create a stronger foundation for responsible classroom integration.

2. AI Literacy Curriculum Support

The IntelliGloss AI Literacy Framework presented in this book can support curriculum development across multiple subject areas. Artificial intelligence literacy is not limited to computer science courses. It intersects with language arts, social studies, science, ethics, and digital citizenship.

District curriculum teams may use this guide to:

- Introduce AI literacy modules within existing courses
- Develop interdisciplinary projects involving AI analysis
- Create student discussions about technology, ethics, and information credibility
- Encourage critical thinking about automated systems

This approach allows schools to integrate AI literacy without replacing established academic subjects.

3. Policy and Academic Integrity Guidance

Many school districts are currently developing policies related to AI-assisted work, academic integrity, and responsible technology use. This book provides a foundation for those discussions.

District leaders may use the content to guide conversations about:

- Acceptable AI use in student assignments
- Disclosure expectations when AI tools assist work
- Teacher guidelines for AI-supported instruction
- Student privacy considerations when using AI platforms

Clear policies help schools balance innovation with accountability.

4. Classroom Implementation

Teachers may use the examples and framework presented in this book to design classroom activities that help students understand artificial intelligence critically rather than simply consume it.

Classroom applications may include:

- Evaluating AI-generated writing
- Comparing AI responses with human research
- Identifying bias or inaccuracies in automated outputs
- Reflecting on the role of technology in modern society

These activities strengthen reasoning skills while promoting responsible use.

5. Building Long-Term AI Literacy

Artificial intelligence will continue evolving, and new tools will emerge regularly. Rather than focusing on specific platforms, this book emphasizes **durable principles and competencies** that remain relevant as technology changes.

By focusing on:

- Conceptual understanding
- Responsible use
- Critical evaluation • Creative application

schools can build a sustainable approach to AI literacy that prepares students for future developments.

Moving from Reaction to Leadership

Educational institutions have historically adapted to major technological shifts. Artificial intelligence represents another important moment of transition.

Schools that approach AI thoughtfully can move beyond reacting to technology and instead **lead students in understanding it responsibly.**

This book is designed to support that leadership by providing a clear framework, practical strategies, and a structured path for integrating artificial intelligence into education with confidence and responsibility.

About the Author
Susie Hala
Founder of the IntelliGloss AI Education Series

Susie Hala is an author, educator, and curriculum developer dedicated to advancing artificial intelligence literacy in education. Her work focuses on helping schools, educators, and students understand how intelligent technologies are shaping modern society—and why that understanding is essential.

Hala is the creator of the **IntelliGloss AI Education Series**, a comprehensive and growing collection of educational books designed to make complex artificial intelligence concepts accessible, structured, and practical for classroom use. Her work emphasizes clarity, organization, and responsible implementation, enabling schools to introduce AI literacy without requiring advanced technical backgrounds.

In addition to her work in education, Hala brings practical, real-world experience working with artificial intelligence systems. Her background includes early exposure to rule-based AI systems, as well as building modern digital platforms that integrate AI tools, automation, and online services. She has developed e-commerce systems and AI-driven solutions, connecting technologies such as payment platforms, hosting environments, and conversational AI tools to create functional, real-world applications.

This combination of foundational understanding and hands-on experience allows her to translate complex AI concepts into language that educators and students can understand and apply.

Recognizing that artificial intelligence is rapidly transforming how information is created, interpreted, and used, Hala advocates for equipping students with the ability not only to use AI tools, but to understand, question, and evaluate them. Her books emphasize conceptual understanding, critical thinking, ethical awareness, and responsible digital citizenship.

The IntelliGloss AI Education Series supports schools in teaching key areas such as artificial intelligence foundations, algorithms and machine learning concepts, AI ethics, and the full spectrum of artificial intelligence development—including Artificial Narrow Intelligence (ANI), Artificial General Intelligence (AGI), and Artificial Superintelligence (ASI)—as well as the evolving role of AI across industries and society. These resources are designed for students in grades 6–12 and for educators seeking structured, classroom-ready approaches to AI literacy.

In addition to her work in education publishing, Hala has a background in entrepreneurship and has spent many years building businesses and developing educational resources that support learning and community growth. Her experience working across multiple industries has shaped her belief that technology education must remain grounded in human values, critical thinking, and ethical responsibility.

Through her writing, Hala seeks to help schools prepare students for a future in which artificial intelligence will influence nearly every aspect of life.

Her goal is clear: to ensure that the next generation understands intelligent technology— not just how to use it, but how it truly works.

Human Values Family

Definition

The Human Values Family in artificial intelligence represents the guiding principles that shape how AI systems are created, used, and integrated into society. These values include ethics, responsibility, trust, and human well-being.

Artificial intelligence may be built on data, algorithms, and computational systems, but its true impact is measured by how it affects people. For this reason, AI is not only a technical system—it is a human-centered system.

Key Concepts

The Human Values Family exists to ensure that artificial intelligence serves humanity in a positive and responsible way. While AI systems can process information and generate outcomes, they do not possess judgment, empathy, or moral understanding.

This means that human values must guide every stage of AI development—from design to deployment.

AI reflects the intentions, decisions, and assumptions of the people who build and use it. If those values are not carefully considered, AI systems can unintentionally produce biased, harmful, or misleading results.

Understanding human values in AI is not optional—it is essential.

Core Elements of the Human Values Family

Ethics
Ethics in AI refers to making decisions that are fair, just, and respectful of human rights. This includes minimizing bias, protecting privacy, and ensuring transparency in how systems operate.

Responsibility
Responsibility means that humans remain accountable for AI systems. Developers, educators, organizations, and users must take ownership of how AI is used and the outcomes it produces.

Trust
Trust is critical for the successful adoption of AI. People must feel confident that AI systems are reliable, safe, and aligned with their best interests.

Human Well-Being
AI should enhance human life—not replace, harm, or diminish it. This includes supporting learning, improving access to information, and creating opportunities while maintaining human dignity.

Real-World Connection

Artificial intelligence is already influencing how people learn, communicate, and make decisions. From search engines to classroom tools, AI is shaping daily experiences.

Without clear human values guiding these systems, there is a risk of reinforcing misinformation, bias, or unrealistic expectations.

For students, this means learning not only how to use AI tools, but how to question them, evaluate their outputs, and understand their limitations.

Applications in Education

In the classroom, the Human Values Family helps students develop critical thinking and ethical awareness. Teachers can use this framework to guide discussions about fairness, responsibility, and the impact of technology on society.

Students can explore questions such as:

How should AI be used responsibly?

What makes an AI system fair?

How does technology influence human behavior?

This approach supports both technical understanding and responsible decision-making.

Benefits

Teaching human values in AI helps students:

Develop ethical awareness

Strengthen critical thinking

Understand the societal impact of technology

Become responsible users and future creators of AI

It ensures that learning about AI goes beyond functionality and includes responsibility.

Challenges

Human values are not always universal. Different cultures, communities, and individuals may have different perspectives on what is considered fair or ethical.

Additionally, AI technology is evolving rapidly, often faster than policies and guidelines can keep up. This makes it essential for educators and students to stay informed and adaptable.

Summary

The Human Values Family reminds us that artificial intelligence is not just about machines—it is about people.

By focusing on ethics, responsibility, trust, and human well-being, students can learn to use AI thoughtfully and responsibly.

Understanding these principles helps prepare learners to navigate a world where artificial intelligence plays an increasingly important role in everyday life.

Table of Contents

Chapter 1
Common Teacher Concerns in the Age of Artificial Intelligence Pg 62

Chapter 2 What Artificial Intelligence Actually Is Pg...66

Chapter 3 Why AI Literacy Matters in K–12 Education Pg 69

Chapter 4 The IntelliGloss AI Literacy Framework Pg 73

Chapter 5 Designing AI-Smart Assignments Pg 77

Chapter 6 AI-Enhanced Lesson Planning for Teachers Pg...81

Chapter 7

Sample Classroom AI Policy and Parent Communication Framework Pg. 83

Chapter 8
A 12-Week AI Literacy Implementation Model for Grades 6–12 Pg. 87

Chapter 9
Assessment Rubrics and Evaluation Tools for AI Literacy Pg. 90

Chapter 10
Assessment and Evaluation in the AI-Ready Classroom Pg. 94

Chapter 11
Standards Alignment and Curriculum Integration Pg. 95

Chapter 12
Institutional Leadership and Administrative Guidance Pg. 96

Chapter 13
Teaching Forward: Human Leadership in an AI-Influenced Era Pg. 99

Chapter 14
The AI Knowledge Ladder: From Bits to Artificial Intelligence Pg. 100 **Chapter 15**
Understanding Bits, Bytes, Data, and AI Training Pg. 104

Chapter 16
The Complete AI School Model: Combining AI-Powered Learning with AI Understanding Pg. 108

Connecting AI Understanding to Cybersecurity Awareness

Definition

Understanding artificial intelligence is the first step toward understanding cybersecurity. While cybersecurity focuses on protecting systems, data, and users, artificial intelligence explains how those systems are built, how they function, and where they may be vulnerable.

The Connection

Every AI system is made up of components such as data, algorithms, models, and infrastructure. These same components are also the targets of cybersecurity threats.

When students understand:

How data is collected and stored

How algorithms make decisions

How AI models process information

How systems operate across frontend and backend layers

They begin to recognize where risks can exist and how systems can be misused.

Why This Matters

In today's digital world, students are not just users of technology—they are participants in complex systems powered by artificial intelligence.

Without understanding how these systems work:

Technology becomes something they trust without question

Risks become harder to recognize

Decisions are made without awareness of consequences

With understanding:

Students become more aware of how their data is used

They recognize how systems can be manipulated

They make more informed and responsible choices

Building Digital Readiness

Digital readiness is not only about using tools. It is about understanding the systems behind those tools.

By learning how AI works, students develop:

Awareness of digital environments

Confidence in navigating technology

A foundation for future learning, including cybersecurity

Conclusion

Cybersecurity begins with understanding.

Before students can protect systems, they must first understand how those systems are built and how they operate. This is why learning artificial intelligence is not only about innovation—it is also about responsibility.

The Foundation of Artificial Intelligence

Definition

The foundation of Artificial Intelligence (AI) refers to the core building blocks that allow AI systems to function. These building blocks work together to help machines learn from information, recognize patterns, and make decisions. AI is not magic—it is a system built on structured components that operate behind the scenes.

Understanding the Foundation in Simple Terms

Artificial Intelligence can be compared to a house. Before a house is complete, it must be built on a strong foundation. This foundation includes concrete, wiring, plumbing, and support structures that are not always visible but are essential for the house to function properly.

In the same way, AI systems rely on foundational components that are not visible to users but are necessary for AI tools to work. When people use AI applications such as chatbots or design tools, they are interacting with the surface. The true power of AI comes from what is underneath.

Key Components of the AI Foundation

1. Data (The Information)

Data is the starting point of all AI systems. It includes text, images, numbers, audio, and other forms of information. AI systems learn by analyzing patterns within this data.

2. Algorithms (The Instructions)

Algorithms are step-by-step instructions that tell the AI system how to process data. They guide how the system learns, identifies patterns, and produces results.

3. Models (The Trained System)

A model is created when data and algorithms are combined through a training process. The model acts as the "brain" of the AI system, allowing it to make predictions and generate responses.

4. Computing Power (The Engine)

AI systems require powerful computers to process large amounts of data. This includes specialized hardware such as processors and graphics processing units (GPUs) that allow AI to function efficiently.

5. Infrastructure (The Environment)

Infrastructure refers to the systems that support AI operations, including servers, storage, networks, and energy systems. Data centers are a key part of this infrastructure, providing the physical space where AI systems operate.

How These Components Work Together

AI systems function by combining all foundational components into a single process. Data is collected and analyzed using algorithms. This process creates a model that can recognize patterns and make decisions. The entire system is powered by computing machines and supported by infrastructure.

Why Understanding the Foundation Matters

Many people use AI tools without understanding how they work. This can lead to confusion, misuse, or over-reliance on technology. By understanding the foundation of AI, students and educators can:

Develop critical thinking skills

Use AI tools more responsibly

Recognize the limitations of AI systems

Build a deeper understanding of modern technology

Real-World Example

When a student asks a chatbot a question, the response does not come from "thinking" like a human. Instead, the AI system uses a trained model that has learned patterns from large amounts of data. It follows algorithms to predict the most appropriate answer, using powerful computers and infrastructure to deliver the response quickly.

This process is also influenced by how the question is asked. The words, structure, and clarity of a question—often referred to as a *prompt*—can affect the quality of the response. Learning how to communicate effectively with AI systems is an important skill known as prompt engineering. For further understanding, students and educators may explore *What to Say When You Talk to Yourself: A Guide to Prompt Engineering* by Susie Hala, which explains how carefully structured prompts can guide AI responses more effectively.

Conclusion

The foundation of Artificial Intelligence is built on data, algorithms, models, computing power, and infrastructure. These components work together to create systems that can learn, predict, and assist humans. Understanding this foundation helps students move beyond simply using AI tools and toward truly understanding how artificial intelligence works.

The Foundational Families of Artificial Intelligence

Introduction

Artificial intelligence is often described as a powerful technology, but in reality, it is not a single system. It is a combination of multiple components working together to create intelligent behavior.

To better understand how artificial intelligence works, it is helpful to break it down into organized groups. In the IntelliGloss framework, these groups are called **Foundational Families of Artificial Intelligence**.

Each family represents a key part of how AI systems are built and operate. Some families focus on how information is created and stored. Others focus on how machines learn, make decisions, process language, or perform tasks. Additional families represent the physical hardware, energy systems, and global infrastructure that make artificial intelligence possible.

When these families work together, they form a complete AI system.

This approach allows students and educators to move beyond simply using artificial intelligence tools and begin to understand what is happening behind the scenes. Instead of viewing AI as a mystery, learners can see it as a structured system made up of clearly defined components.

The diagram on the next page provides a visual overview of these families and how they connect to form the foundation of artificial intelligence.

By learning each family step by step, students develop a deeper understanding of how AI systems process information, learn from data, and interact with the world.

Understanding the Foundational Families of Artificial Intelligence is the first step toward true AI literacy.

The Foundational Families of Artificial Intelligence

Family	Role	Description
Bit Family	Digital Building Blocks	The smallest unit of digital information (0 or 1). All computing systems begin with bits.
Algorithm Family	Reasoning Engine	Step-by-step instructions that tell computers how to solve problems and process information.
Token Family	Language Units	Pieces of text used by AI models to understand and generate human language.

Family	Role	Description
Parameter Family	Learning Adjustment System	Adjustable numerical values inside AI models that allow them to learn patterns from data.
Data Family	Learning Material	The information AI studies in order to learn patterns, relationships, and knowledge.
Logic Family	Decision Rules	The reasoning structures that allow AI systems to make decisions and evaluate conditions.
Memory Family	Information Storage	Systems that store and retrieve information so AI models can access past knowledge and context.
Knowledge Family	Understanding & Meaning	Structured information that allows AI to represent facts, relationships, and concepts.
Neural Network Family	Learning Brain	Interconnected layers of artificial neurons that enable AI systems to recognize patterns, learn from data, and make predictions.
AI Model Family	Intelligence Engine	The trained system that uses data, parameters, and neural networks to perform tasks such as prediction, classification, and generation.
AI Translator Architecture Family	Language Conversion System	Systems that transform input into output, enabling AI to convert text, speech, or images into meaningful responses.
AI Agent Family	Action and Task Execution System	AI systems designed to perceive information, make decisions, and carry out tasks autonomously or semi-autonomously.
Chip Family	Hardware Brain	Physical processors (GPUs, TPUs, CPUs, NPUs) that perform AI computations.
Watts Family	Energy Power	The electrical power required to run AI hardware and data centers.
AI Energy Family	Energy Source System	The total energy consumed over time to operate AI systems, including training and real-time usage.
Infrastructure Family	Global Support System	Data centers, internet networks, cloud systems, and global infrastructure that allow AI to operate and scale.
Ingredient Family	System Composition	The essential components that come together to create an AI system, including data, algorithms, models, hardware, and energy.
Qubit Family	Quantum Information Units	Quantum bits that can represent multiple states simultaneously, forming the foundation of quantum computing.

The Bit Family

Introduction

The Bit Family represents the starting point of all digital technology. Every computer system, application, and artificial intelligence model is built from bits—the smallest unit of information in computing.

A bit can hold one of two values: 0 or 1. While this may seem simple, these two values form the foundation of everything digital. When bits are combined and organized, they create larger units of data that allow computers to store text, display images, play videos, and run complex systems.

From a single bit to massive data systems, all digital information follows this same structure. This is how computers are able to represent and process the world in a form they can understand.

In artificial intelligence, bits are essential because they store the data that models learn from, process the calculations that drive decision-making, and support the systems that generate responses.

Understanding the Bit Family helps students see how simple binary signals grow into powerful technologies. It reveals that behind every advanced AI system is a foundation built from the most basic building blocks of information.

Bit Family — Visual Understanding of Data Size

How Much Data Can Each Unit Hold? (Simple Examples)

Unit	Size	Photos (Approx.)	Video (Approx. Hours)
Bit	1 bit	N/A	N/A
Byte	8 bits	1 character	N/A
Kilobyte (KB)	~1,000 bytes	Part of a paragraph	N/A
Megabyte (MB)	~1,000 KB	~1 photo	~1 minute of video
Gigabyte (GB)	~1,000 MB	~250 photos	~1–2 hours of video
Terabyte (TB)	~1,000 GB	~250,000 photos	~1,000–2,000 hours of video
Petabyte (PB)	~1,000 TB	~250 million photos	~1–2 million hours of video
Exabyte (EB)	~1,000 PB	~250 billion photos	~1–2 billion hours of video
Zettabyte (ZB)	~1,000 EB	~250 trillion photos	~1–2 trillion hours of video
Yottabyte (YB)	~1,000 ZB	~250 quadrillion photos	~1–2 quadrillion hours of video

"Learn AI. Understand AI. Shape the Future."
— Susie Hala

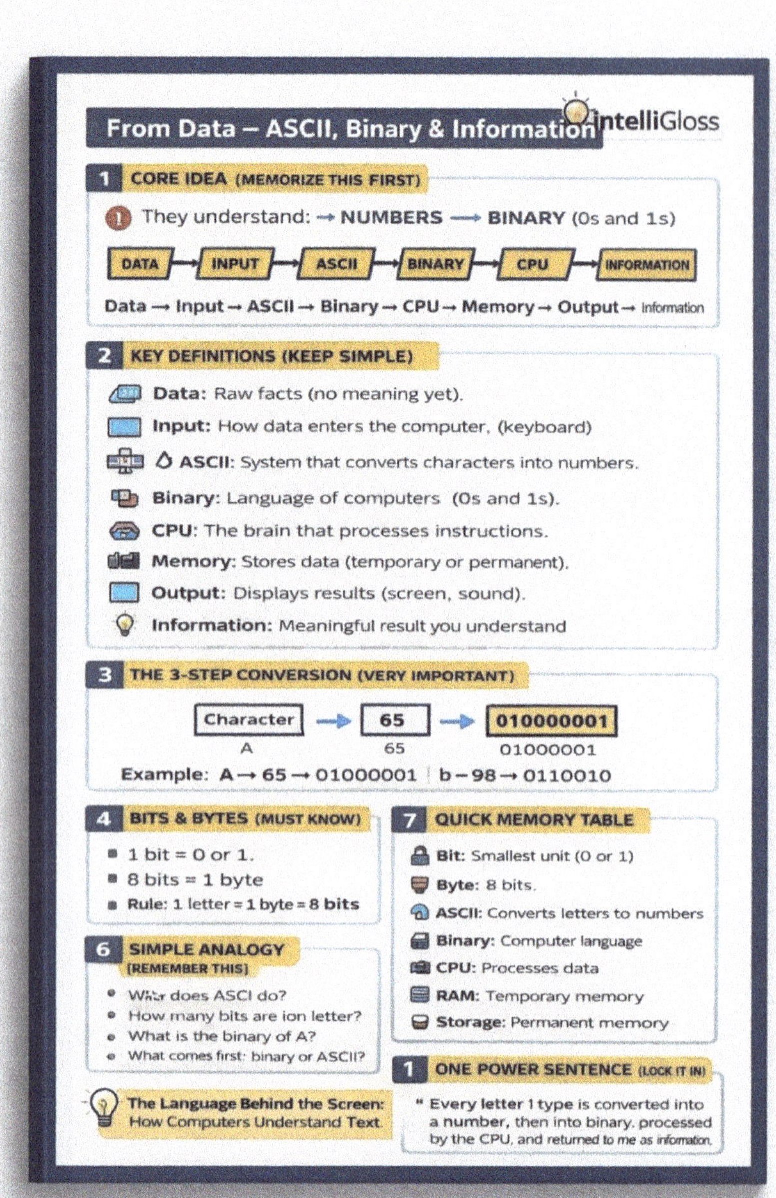

The Algorithm Family

Definition and Overview

The **Algorithm Family** represents the *thinking engine* of artificial intelligence — the set of mathematical instructions that guide how AI analyzes data, learns patterns, and makes decisions.

An **algorithm** is a sequence of steps designed to solve a problem or complete a task. In AI, algorithms tell the system *how to learn*, *how to predict*, and *how to improve* with every iteration.

In simple terms:

The Algorithm Family is the *logic of intelligence* — the invisible code that transforms data into decisions and predictions into progress.

How Algorithms Power AI

Every AI system — from simple chatbots to large-scale deep learning models — relies on algorithms as its foundation.
They determine how the model:

Processes and cleans data,

Identifies relationships between variables,

Learns from feedback, and

Improves over time.

Without algorithms, even the most powerful dataset or neural network would remain static and lifeless.

In essence, **algorithms are the teachers** that train AI how to think.

⚙️ *Why the Algorithm Family Matters*

Algorithms are what separate automation from intelligence.
They allow machines to **reason**, **adapt**, and **learn** — not just follow commands.
For prompt engineers, understanding algorithmic behavior helps explain *why* AI sometimes misunderstands a question, repeats patterns, or changes tone after feedback.

Every AI response you receive follows an underlying process — a pattern recognition journey from input to output.

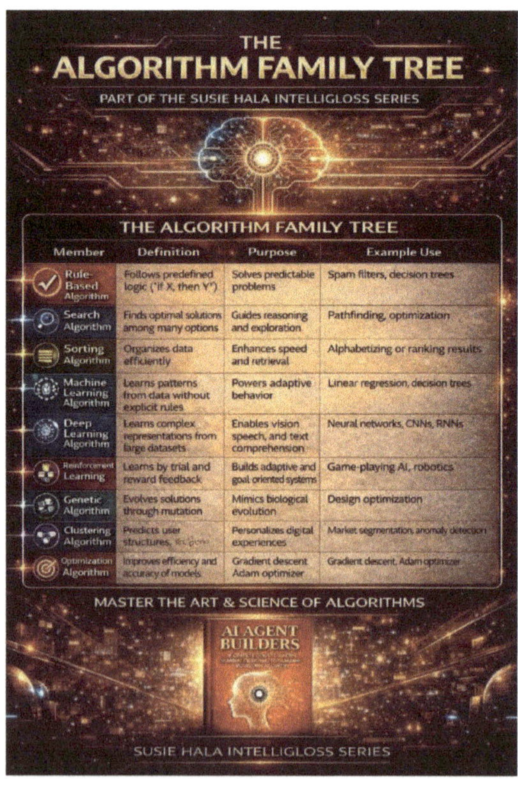

Member	Definition	Purpose	Example Use
Rule-Based Algorithm	Follows predefined logic ("if X, then Y")	Solves predictable problems	Spam filters, decision trees
Search Algorithm	Finds optimal solutions among many options	Guides reasoning and exploration	Pathfinding, optimization
Sorting Algorithm	Organizes data efficiently	Enhances speed and retrieval	Alphabetizing or ranking results
Machine Learning Algorithm	Learns patterns from data without explicit rules	Powers adaptive behavior	Linear regression, decision trees
Deep Learning Algorithm	Learns complex representations from large datasets	Enables vision, speech, and text comprehension	Neural networks, CNNs, RNNs
Reinforcement Learning	Learns by trial and reward feedback	Builds adaptive and goal oriented systems	Game-playing AI, robotics
Genetic Algorithm	Evolves solutions through mutation	Mimics biological evolution	Design optimization
Clustering Algorithm	Predicts user structures, regions	Personalizes digital experiences	Market segmentation, anomaly detection
Optimization Algorithm	Improves efficiency and accuracy of models	Gradient descent, Adam optimizer	Gradient descent, Adam optimizer

Analogy for Everyday Readers

Imagine AI as a chef in a kitchen:

The **Dataset Family** provides the ingredients.

The **Knowledge Family** gives the recipes.

The **Algorithm Family** is the *cooking process* — mixing, heating, tasting, and adjusting until the dish is perfect.

Just as cooking methods determine the quality of food, algorithms determine the quality of intelligence.

Algorithm Efficiency and Energy Use

Not all algorithms are equal.
Some are **fast but shallow**, while others are **deep but energy-intensive**.
Training large AI models requires algorithms that can handle billions of calculations per second — powered by the **Watts Family** (energy) and executed through the **Chip Family** (hardware).

The design of efficient algorithms directly affects sustainability in AI — minimizing power consumption while maximizing accuracy and speed.

Algorithms and Prompt Engineering

For prompt engineers, algorithms explain *how* the AI arrives at its answers.
When a model gives inconsistent or off-topic results, it's often due to:

Misinterpretation of input signals (input encoding),

Bias in training algorithms, or

Overfitting to certain data patterns.

Knowing this helps you craft prompts that guide the AI's decision-making — not just its output.

Quick Takeaway: The Algorithm Family is the reasoning heart of AI.
It transforms information into understanding and enables machines to think, learn, and evolve like digital problem-solvers.

Token Family Introduction

The Token Family represents one of the most important bridges between human language and artificial intelligence. While humans communicate using words, sentences, and meaning, AI systems do not understand language in the same way. Instead, they rely on tokens—small units of text that allow machines to process, analyze, and generate language step by step.

A token can be a word, part of a word, a character, or even punctuation. When a student types a sentence into an AI system, that sentence is not read as a whole idea. It is first broken down into tokens. These tokens become the input that the AI model uses to recognize patterns, predict outcomes, and generate responses.

The Token Family works closely with several other foundational families in artificial intelligence. It connects directly with the Data Family, which provides the text used for training, and the Neural Network Family, which processes token patterns to produce intelligent outputs. It also plays a key role in the AI Translator Architecture Family, where tokens are transformed, analyzed, and reassembled into meaningful responses.

Understanding tokens helps students see what is happening behind the scenes when they interact with AI tools. Instead of viewing AI as a "magic system," they begin to understand that every response is built from sequences of tokens processed through mathematical models.

In simple terms, if language is what humans speak, tokens are what AI understands.

Token Family Breakdown Chart

Component	Role	Description	Example
Tokenization	Input Conversion	The process of breaking text into smaller units (tokens) that AI can process	"Artificial Intelligence" → "Artificial" + "Intelligence"
Tokens	Language Units	The individual pieces of text used by AI models for understanding and generation	"AI", "learn", "ing", "."
Subword Tokens	Efficiency Units	Words split into smaller parts to handle unknown or complex vocabulary	"unbelievable" → "un", "believ", "able"
Character Tokens	Fine-Grained Units	Text broken down into individual characters for detailed processing	"AI" → "A" + "I"
Vocabulary (Token Set)	Token Library	The complete set of tokens an AI model recognizes and uses	GPT models have thousands to millions of tokens
Encoding	Token Mapping	Converting tokens into numerical representations for computation	"AI" → [1234, 5678] (example IDs)
Decoding	Output Reconstruction	Converting numerical outputs back into readable text	[1234, 5678] → "AI system"
Context Window	Memory Limit	The number of tokens an AI can process at one time	Example: 8,000 tokens ≈ several pages of text
Token Embeddings	Meaning Representation	Mathematical vectors that capture relationships between tokens	"king" and "queen" have similar embeddings
Token Prediction	Language Generation	The process of predicting the next token in a sequence	"AI is" → predicts "powerful"

The Token Family reminds us that behind every intelligent response is a structured sequence of tokens working together. What feels like natural conversation to humans is, for AI, a carefully processed stream of language units transformed into meaning.

The AI Ecosystem Family

The AI Ecosystem Family is a connected group of artificial intelligence technologies, systems, behaviors, and applications that work together to help machines perform intelligent tasks. This ecosystem includes core AI systems, AI agents, agentic AI, AI autonomy, autonomous AI workers, and generative AI. Each part plays a different role in how modern AI systems think, learn, create, make decisions, solve problems, and interact with the world.

At the center of the ecosystem is artificial intelligence itself — the broad field focused on building machines and software that can perform tasks that normally require human intelligence. AI agents act as the systems that carry out goals and actions. Agentic AI describes the ability of AI systems to plan, reason, adapt, and pursue objectives with limited human guidance. AI autonomy represents the level of independence an AI system has while making decisions and taking actions. AI autonomous workers apply these capabilities to perform real-world tasks like digital employees or assistants. Generative AI adds the ability to create new content such as text, images, music, video, code, and designs.

The AI Ecosystem Family demonstrates that modern artificial intelligence is no longer just one single technology. Instead, it is an interconnected environment of systems, tools, intelligence, automation, creativity, and human collaboration. As AI continues to evolve, these technologies are becoming increasingly integrated into education, business, healthcare, communication, cybersecurity, research, and everyday life.

Understanding the AI Ecosystem Family helps students, educators, and readers see how the different parts of AI connect together to shape the future of intelligent systems and the digital world.

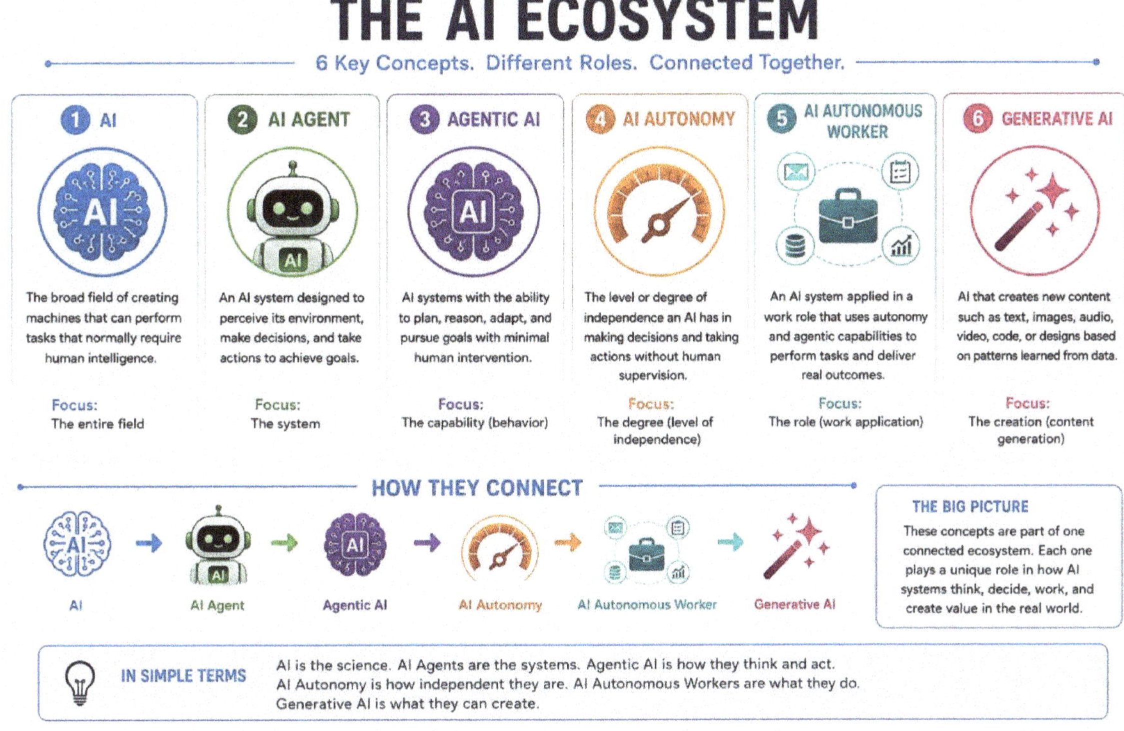

Parameter Family Introduction

The Parameter Family represents the internal learning system of artificial intelligence. While tokens provide the language and data provides the learning material, parameters are what allow AI systems to adjust, improve, and make intelligent decisions over time.

In simple terms, parameters are numerical values inside an AI model that change as the system learns. These values are not visible to users, but they play a critical role in how the model understands patterns, relationships, and meaning. Every time an AI system is trained, it adjusts its parameters to better predict outcomes and generate accurate responses.

Parameters exist throughout neural networks, where they act as the connections between artificial neurons. Each connection has a weight, which determines how important a particular piece of information is. During training, these weights are continuously adjusted based on data, allowing the system to improve its performance.

The Parameter Family works closely with the Data Family, which provides the information needed for learning, and the Neural Network Family, which organizes how parameters are structured and applied. It also supports the Token Family by helping the model interpret and predict sequences of language.

Understanding parameters helps students recognize that AI is not simply programmed with fixed rules. Instead, it learns by adjusting internal values that guide how it processes information. These adjustments are what allow AI systems to evolve from simple pattern recognition to complex reasoning and decision-making.

In essence, if data is what AI learns from, parameters are how AI learns.

Parameter Family Breakdown Chart

Component	Role	Description	Example
Parameters	Learning Values	Numerical values inside an AI model that are adjusted during training	Millions or billions of values inside a model
Weights	Connection Strength	Values that determine how strongly one neuron influences another	A higher weight = stronger influence
Bias	Adjustment Factor	A value added to help the model shift and fine-tune outputs	Helps improve accuracy in predictions
Training Process	Learning Mechanism	The process of adjusting parameters using data to improve performance	Model improves after analyzing many examples

Component	Role	Description	Example
Loss Function	Error Measurement	Measures how far the model's prediction is from the correct answer	Lower loss = better performance
Optimization Algorithm	Adjustment Strategy	Method used to update parameters efficiently	Gradient Descent adjusts weights step by step
Gradient	Direction of Change	Indicates how parameters should change to reduce error	Shows whether to increase or decrease a value
Backpropagation	Learning Process	System that sends error signals backward to update parameters	Adjusts earlier layers based on output error
Hyperparameters	Control Settings	External settings that guide how training happens (not learned)	Learning rate, batch size
Model Size	Capacity Indicator	The total number of parameters in a model	Larger models = more learning capacity

The Parameter Family is the hidden engine of learning in artificial intelligence. While users interact with outputs, it is the continuous adjustment of parameters behind the scenes that makes intelligent behavior possible.

Data Family Introduction

The Data Family represents the foundation of learning in artificial intelligence. Every AI system depends on data to develop knowledge, recognize patterns, and make decisions. Without data, artificial intelligence cannot learn, adapt, or function effectively.

Data can take many forms, including text, images, audio, video, and numerical information. These inputs provide the raw material that AI systems analyze during training. By studying large amounts of data, AI models begin to identify patterns, relationships, and structures that allow them to make predictions and generate responses.

The quality and diversity of data play a critical role in how well an AI system performs. Accurate, balanced, and relevant data helps models produce reliable results, while poor or biased data can lead to incorrect or unfair outcomes. This is why data collection, preparation, and evaluation are essential steps in the development of responsible AI systems.

The Data Family works closely with the Parameter Family, which adjusts internal values based on data, and the Neural Network Family, which processes and organizes data through layers of computation. It also supports the Token Family when dealing with language, as text data is converted into tokens for analysis.

Understanding the Data Family helps students recognize that AI does not "know" things on its own. Instead, it learns from the information it is given. The more meaningful and well-prepared the data, the more capable the AI system becomes.

In simple terms, data is the source of knowledge, and it is the starting point for all artificial intelligence.

Data Family Breakdown Chart

Component	Role	Description	Example
Data	Learning Material	Raw information used by AI systems to learn patterns and relationships	Text, images, audio, numbers
Dataset	Organized Collection	A structured group of data used for training and evaluation	A folder of labeled images
Training Data	Learning Input	Data used to teach the AI model during training	Thousands of sentences for a language model
Validation Data	Performance Check	Data used to tune the model during training without directly learning from it	Helps prevent overfitting
Test Data	Final Evaluation	Data used to measure how well the model performs after training	New unseen examples
Structured Data	Organized Format	Data arranged in tables with clear categories	Spreadsheets, databases
Unstructured Data	Flexible Format	Data without a predefined structure	Text documents, images, videos
Labeled Data	Guided Learning	Data tagged with correct answers to guide training	Image labeled "cat" or "dog"
Unlabeled Data	Self-Learning Input	Data without labels, used in unsupervised learning	Raw text without categories
Data Quality	Accuracy Measure	The reliability and correctness of data	Clean, complete, and consistent data
Data Bias	Imbalance Issue	When data does not represent all groups fairly	Skewed or incomplete datasets
Data Preprocessing	Preparation Step	Cleaning and organizing data before training	Removing errors, formatting text

The Data Family reminds us that artificial intelligence is only as strong as the information it learns from. High-quality data leads to meaningful insights, while poor data can limit or misguide intelligent systems.

Memory Family Introduction

The Memory Family represents how artificial intelligence systems store, retain, and retrieve information. Just as humans rely on memory to recall past experiences, learn new concepts, and make decisions, AI systems depend on memory to access previously processed information and maintain context.

In artificial intelligence, memory is not a single component but a collection of mechanisms that allow systems to handle information over time. Some forms of memory are short-term, holding information temporarily while a task is being completed. Other forms are long-term, storing learned knowledge that can be used across different tasks and interactions.

Memory plays a critical role in enabling AI systems to understand sequences, maintain context in conversations, and improve decision-making. For example, when a user interacts with a language model, the system uses memory to keep track of previous words, sentences, or prompts in order to generate coherent and relevant responses.

The Memory Family works closely with the Data Family, which provides the information to be stored, and the Parameter Family, which encodes learned knowledge within the model. It also supports the Neural Network Family by enabling systems to process sequences and retain important patterns over time.

Understanding the Memory Family helps students see that AI is not simply reacting to individual inputs in isolation. Instead, it relies on stored information and contextual awareness to produce meaningful and consistent outputs.

In simple terms, if data is what AI learns from, memory is how AI remembers and uses what it has learned.

Memory Family Breakdown Chart

Component	Role	Description	Example
Memory	Information Storage	The ability of an AI system to store and access information	Retaining previous inputs in a task
Short-Term Memory	Temporary Storage	Holds information for immediate processing	Keeping track of words in a sentence
Long-Term Memory	Persistent Storage	Stores learned knowledge for future use	Knowledge embedded in model parameters
Context Memory	Conversation Tracking	Maintains information within a specific interaction	Remembering earlier parts of a prompt
Working Memory	Active Processing	Handles information currently being used in computation	Processing a sequence step by step

Component	Role	Description	Example
External Memory	Extended Storage	Memory stored outside the model for retrieval	Databases, vector stores
Internal Memory	Embedded Knowledge	Information stored within the model itself	Learned patterns inside neural networks
Sequence Memory	Order Awareness	Tracks the order of inputs over time	Understanding sentence structure
Retrieval Mechanism	Access System	Retrieves stored information when needed	Searching relevant past data
Attention Mechanism	Focus System	Highlights important parts of memory for processing	Focusing on key words in a sentence
Memory Capacity	Storage Limit	The amount of information the system can handle at once	Context window size

The Memory Family shows that intelligence is not just about processing information in the moment, but about retaining, organizing, and using knowledge over time to create meaningful understanding.

"Learn AI. Understand AI. Shape the Future."
— Susie Hala

Knowledge Family Introduction

The Knowledge Family represents how artificial intelligence organizes, connects, and uses information to create understanding. While data provides raw information and memory stores it, knowledge is what gives that information meaning and structure.

In artificial intelligence, knowledge is not simply a collection of facts. It is the result of identifying relationships, patterns, and connections within data. AI systems use knowledge to recognize concepts, make decisions, and generate meaningful responses. This allows machines to move beyond simple data processing and toward more intelligent behavior.

Knowledge can be represented in different ways within AI systems. Some forms are structured, such as databases and knowledge graphs, where relationships between concepts are clearly defined. Other forms are learned implicitly through neural networks, where patterns are encoded within parameters and used to guide predictions.

The Knowledge Family works closely with the Data Family, which provides the information needed to build knowledge, and the Memory Family, which stores and retrieves that information. It also connects with the Parameter Family, where learned knowledge is embedded within the model, and the Neural Network Family, which processes and organizes that knowledge.

Understanding the Knowledge Family helps students recognize that AI systems do not simply memorize information. Instead, they learn how pieces of information relate to one another, allowing them to interpret meaning, solve problems, and respond intelligently.

In simple terms, if data is information and memory stores it, knowledge is the understanding that makes it useful.

Knowledge Family Breakdown Chart

Component	Role	Description	Example
Knowledge	Understanding System	Organized information that allows AI to interpret meaning and make decisions	Recognizing relationships between concepts
Facts	Basic Information	Individual pieces of information stored within a system	"Water freezes at 0°C"
Relationships	Connection Mapping	Links between pieces of information that create meaning	"Teacher teaches student"
Knowledge Representation	Structure System	Methods used to organize and store knowledge in AI	Knowledge graphs, semantic networks
Knowledge Graph	Network Mapping	A structured representation of entities and their relationships	Google Knowledge Graph
Semantic Understanding	Meaning Interpretation	The ability to understand meaning behind words and data	Understanding synonyms and context
Inference	Reasoning Process	Drawing conclusions based on known information	If A = B and B = C, then A = C
Rules	Decision Framework	Logical guidelines used to make decisions	If temperature < 0°C → freeze
Ontology	Concept Framework	A structured system defining categories and relationships	Classification of animals
Explicit Knowledge	Direct Information	Clearly defined and stored knowledge	Rules, databases
Implicit Knowledge	Learned Patterns	Knowledge learned through experience and data	Patterns learned by neural networks
Knowledge Integration	System Coordination	Combining information from multiple sources	Merging text, images, and data

The Knowledge Family transforms information into understanding, allowing artificial intelligence to move beyond data and memory into meaningful reasoning and intelligent decision-making.

"Learn AI. Understand AI. Shape the Future."
— Susie Hala

AI Model Family Introduction

The AI Model Family represents the core system within artificial intelligence that performs tasks such as prediction, classification, and content generation. It is the part of AI that users interact with, whether they are asking questions, generating images, or analyzing data.

An AI model is a trained system that has learned patterns from data. During training, the model processes large amounts of information and adjusts its internal parameters to improve accuracy. Once trained, the model can take new input and produce meaningful output based on what it has learned.

AI models can take many forms depending on their purpose. Some models are designed to recognize images, others to understand language, and others to make decisions or predictions. Large Language Models (LLMs), for example, are designed to process and generate human language by predicting sequences of tokens.

The AI Model Family works closely with the Data Family, which provides the learning material, and the Parameter Family, which stores learned patterns. It also relies on the Neural Network Family for structure and the Chip and Infrastructure Families for execution and deployment.

Understanding the AI Model Family helps students see that AI is not just a concept, but a working system that has been trained to perform specific tasks. It is the engine that transforms input into output.

In simple terms, if AI is the system, the model is the part that actually does the thinking and producing.

AI Model Family Breakdown Chart

Component	Role	Description	Example
AI Model	Core System	The trained system that performs tasks	Chatbot, image generator
Training	Learning Process	Process of teaching the model using data	Learning from datasets
Inference	Output Generation	Using the trained model to produce results	Answering a question
Model Architecture	Structural Design	Defines how the model is built	Neural networks, transformers
Parameters	Learned Values	Internal values adjusted during training	Billions of weights

Component	Role	Description	Example
Input Data	Entry Information	Data provided to the model for processing	User prompt
Output Data	Result	The response or prediction generated	AI-generated text
Model Types	Functional Categories	Different kinds of models for different tasks	Classification, generation
Fine-Tuning	Model Adjustment	Improving a model for specific tasks	Custom-trained AI
Pretrained Model	Base System	A model trained on large datasets before use	General-purpose AI
Evaluation	Performance Check	Measuring how well the model performs	Accuracy, loss metrics

The AI Model Family represents the working core of artificial intelligence, where learned patterns are transformed into meaningful actions, predictions, and responses.

Neural Network Family Introduction

The Neural Network Family represents the structural foundation of how artificial intelligence learns and processes information. Inspired by the human brain, neural networks are systems of interconnected units, often called neurons, that work together to analyze data, recognize patterns, and make decisions.

In artificial intelligence, a neural network is made up of layers. The input layer receives data, the hidden layers process that data, and the output layer produces a result. As information flows through these layers, the network applies mathematical transformations that allow it to detect patterns and relationships within the data.

Neural networks rely heavily on the Parameter Family, where weights and biases determine how strongly information flows between neurons. During training, these parameters are adjusted to improve the network's performance. The more the network learns, the better it becomes at recognizing patterns and making accurate predictions.

The Neural Network Family works closely with the Data Family, which provides the information needed for learning, and the Memory and Knowledge Families, which help store and organize learned information. It also supports the Token Family when processing language, allowing AI systems to interpret sequences of tokens in meaningful ways.

Understanding neural networks helps students see that AI is not simply following instructions. Instead, it is learning from examples and improving over time through layered processing and pattern recognition.

In simple terms, neural networks are the systems that allow AI to think, learn, and make sense of information.

Neural Network Family Breakdown Chart

Component	Role	Description	Example
Neural Network	Learning Structure	A system of interconnected neurons that processes data and learns patterns	Image recognition system
Neuron (Node)	Processing Unit	A basic unit that receives input, processes it, and passes output forward	A single calculation point
Input Layer	Data Entry Point	The first layer that receives raw data	Pixels of an image
Hidden Layers	Processing Layers	Intermediate layers that transform data and detect patterns	Feature detection layers
Output Layer	Result Generator	The final layer that produces predictions or decisions	"Cat" vs "Dog" classification
Weights	Signal Strength	Values that determine how strongly inputs influence outputs	Higher weight = stronger impact
Bias	Adjustment Value	Helps shift outputs to improve learning accuracy	Fine-tuning predictions
Activation Function	Decision Function	Determines whether a neuron should activate based on input	ReLU, Sigmoid
Forward Propagation	Data Flow	The process of passing input data through the network to produce output	Input → Hidden → Output
Backpropagation	Learning Process	Adjusts weights and biases based on error to improve performance	Correcting mistakes
Loss Function	Error Measurement	Calculates how far the output is from the correct answer	Prediction vs actual
Deep Neural Network	Advanced Structure	A network with many hidden layers for complex learning	Deep learning models

The Neural Network Family provides the structure that allows artificial intelligence to transform data into understanding, making it possible for machines to learn from experience and improve over time.

Dual Language System in Artificial Intelligence

Definition

The Dual Language System in Artificial Intelligence refers to the way AI communicates using two different forms of language at the same time: human language for interaction and machine language for internal processing.

Key Components

Human Language (Natural Language)
This is the language people use to communicate with AI, such as English, Spanish, Tongan, or Fijian. It includes words, sentences, and questions that users type or speak.

Machine Language (Numerical Representation)
This is how AI understands information internally. Words are converted into numbers called tokens and processed using mathematical models and algorithms.

How It Works (Step-by-Step)

A user enters a question in human language

The AI converts the words into numerical tokens

The AI processes the tokens using a neural network

The AI generates a response based on patterns and probabilities

The response is converted back into human language

Example

User Input:
"What is Artificial Intelligence?"

AI Process:

Converts words into numbers

Analyzes patterns from training data

Predicts the best possible answer

Output:
"Artificial Intelligence is the ability of machines to perform tasks that normally require human intelligence."

Why This Matters

Understanding the dual language system helps students realize that:

AI does not "think" like humans

AI translates language into math and back into language

Every interaction with AI is based on patterns, not emotions or understanding

Applications

Chatbots and virtual assistants

Language translation systems

Search engines

Voice recognition tools

Challenges

Misinterpretation of human language

Bias in language data

Lack of true understanding or reasoning

Summary

Artificial Intelligence works by translating human language into machine-readable data and then converting the results back into human language. This dual system allows AI to interact with people while operating through mathematical processes behind the scenes.

AI Translator Architecture Family Introduction

The AI Translator Architecture Family represents the system that transforms input into meaningful output in artificial intelligence. It acts as the bridge between what humans provide and what AI produces. Whether a user types a question, uploads an image, or speaks a command, this architecture is responsible for interpreting that input and generating a response.

In artificial intelligence, translation does not only mean converting one language to another. It refers to the broader process of transforming one form of information into another. For example, text can be translated into a response, speech can be converted into text, and images can be interpreted into descriptions. This transformation process is at the core of how modern AI systems function.

The AI Translator Architecture is built on multiple foundational families. It relies on the Token Family to break input into manageable units, the Neural Network Family to process patterns, and the Parameter Family to guide learning. It also uses knowledge and memory to maintain context and produce coherent results.

At the center of this architecture is the ability to encode and decode information. Encoding transforms input into a format the model can understand, while decoding converts processed information back into human-readable output. This continuous transformation allows AI systems to interact with users in a natural and meaningful way.

Understanding the AI Translator Architecture Family helps students see that AI is not simply responding randomly. Instead, it is systematically translating input into output through structured processes and learned patterns.

In simple terms, if tokens are the language pieces and neural networks are the brain, the AI Translator Architecture is the system that turns understanding into communication.

AI Translator Architecture Family Breakdown Chart

Component	Role	Description	Example
AI Translator Architecture	Transformation System	Converts input into meaningful output	Question → Answer
Input Processing	Data Intake	Receives and prepares user input for analysis	Text prompt entered by a user
Encoding	Input Conversion	Transforms input into numerical representations	Words → token IDs
Context Building	Meaning Formation	Organizes input into a structured understanding	Sentence context tracking
Sequence Processing	Order Handling	Maintains the correct order of information	Word sequence in a sentence
Attention Mechanism	Focus System	Identifies important parts of input data	Highlighting key words

Component	Role	Description	Example
Representation Layer	Feature Mapping	Converts data into internal representations for processing	Embeddings
Decoding	Output Generation	Converts processed data back into human-readable form	Token IDs → words
Output Generation	Response Creation	Produces the final answer or result	AI-generated response
Multimodal Translation	Cross-Format Processing	Translates between different types of data	Image → text description
Feedback Loop	Improvement Cycle	Uses results to refine future outputs	Model learning over time

The AI Translator Architecture Family shows that artificial intelligence is not just about understanding information, but about transforming it into meaningful communication that humans can use and interact with.

AI Architecture Family Introduction

The AI Architecture Family represents the overall design and structure of an artificial intelligence system. While individual families such as Data, Parameters, and Neural Networks explain how AI learns and processes information, the AI Architecture Family shows how all these components are organized and work together as a complete system.

In artificial intelligence, architecture refers to the blueprint that defines how data flows through a system, how components interact, and how decisions are produced. It determines how input is received, how it is processed, and how output is generated. A well-designed architecture allows AI systems to operate efficiently, scale to large problems, and produce reliable results.

AI architectures can vary depending on the type of system being built. Some are designed for image recognition, others for language processing, and others for decision-making or automation. Despite these differences, most architectures share common elements such as input layers, processing layers, memory components, and output mechanisms.

The AI Architecture Family connects all foundational families. It integrates the Data Family for input, the Token Family for language processing, the Neural Network Family for computation, the Parameter Family for learning, and the Memory and Knowledge Families for storing and organizing information. It also works closely with the AI Translator Architecture Family to transform input into meaningful output.

Understanding the AI Architecture Family helps students see the "big picture" of artificial intelligence. Instead of viewing AI as separate parts, they begin to understand it as a coordinated system where each component plays a specific role.

In simple terms, if individual families are the parts of AI, the AI Architecture Family is the blueprint that brings everything together into one working system.

AI Architecture Family Breakdown Chart

Component	Role	Description	Example
AI Architecture	System Blueprint	The overall design that defines how an AI system is structured and operates	ChatGPT system design
Input Layer	Data Entry Point	Receives raw input from users or external sources	Text prompt, image upload
Preprocessing Layer	Data Preparation	Cleans and organizes input data before processing	Tokenization, normalization
Processing Core	Computation Engine	The main system where data is analyzed and transformed	Neural networks, transformers
Parameter System	Learning Core	Stores adjustable values that guide learning and predictions	Model weights and biases
Memory System	Context Storage	Maintains and retrieves information during processing	Context window, external memory
Knowledge System	Understanding Layer	Organizes relationships and meaning within the system	Knowledge graphs, learned patterns
Decision Layer	Output Logic	Determines the final output based on processed data	Selecting the best response
Output Layer	Result Delivery	Presents results to the user in a usable format	Text, image, or audio response
Feedback Loop	Improvement System	Uses outcomes to refine future performance	Model updates and retraining
Scalability System	Expansion Capability	Allows the system to handle increasing data and complexity	Cloud-based AI systems
Integration Layer	System Connection	Connects AI with external tools and platforms	APIs, databases, applications

The AI Architecture Family reveals that artificial intelligence is not a single technology, but a coordinated system of interconnected components working together to transform data into intelligent outcomes.

Chip Family Introduction

The Chip Family represents the physical hardware that powers artificial intelligence systems. While software components such as data, algorithms, and neural networks define how AI works, none of these systems can function without the computing chips that perform the actual calculations.

Chips are specialized electronic components designed to process information at extremely high speeds. In artificial intelligence, these chips handle the complex mathematical operations required for training models and generating outputs. Every prediction, calculation, and response produced by an AI system is executed by hardware within the Chip Family.

Different types of chips are used depending on the needs of the system. Central Processing Units (CPUs) handle general-purpose tasks, while Graphics Processing Units (GPUs) are optimized for parallel processing, making them ideal for training neural networks. More specialized chips, such as Tensor Processing Units (TPUs) and Neural Processing Units (NPUs), are designed specifically for AI workloads, enabling faster and more efficient computation.

The Chip Family works closely with the Neural Network Family, which defines the structure of computation, and the Parameter Family, which stores the values being processed. It also connects to the Infrastructure Family, where large-scale systems of chips are organized in data centers and cloud environments.

Understanding the Chip Family helps students recognize that artificial intelligence is not purely abstract. It depends on physical machines that require power, resources, and engineering to operate. These chips are the engines that bring AI models to life.

In simple terms, if AI is the intelligence, chips are the machines that make that intelligence possible.

Chip Family Breakdown Chart

Component	Role	Description	Example
Chip (Processor)	Computation Engine	Electronic hardware that performs calculations required for AI	CPU, GPU
CPU (Central Processing Unit)	General Processor	Handles a wide range of computing tasks	Running basic applications

Component	Role	Description	Example
GPU (Graphics Processing Unit)	Parallel Processor	Processes many calculations at once, ideal for AI training	Training neural networks
TPU (Tensor Processing Unit)	AI Accelerator	Specialized chip designed for machine learning tasks	Google TPU
NPU (Neural Processing Unit)	Edge AI Processor	Optimized for running AI on devices like phones	Smartphone AI features
AI Accelerator	Performance Booster	Hardware designed to speed up AI computations	Dedicated AI chips
Memory (Hardware)	Data Storage	Stores data and instructions for processing	RAM, VRAM
Parallel Processing	Speed Mechanism	Ability to perform multiple calculations simultaneously	GPUs processing thousands of operations
Throughput	Processing Capacity	Amount of data processed over time	High-performance computing systems
Latency	Response Time	Time it takes to process a request	Faster chips = lower latency
Power Consumption	Energy Usage	Amount of energy required to run the chip	High-performance GPUs use more power
Edge Devices	Local Processing	Devices that run AI locally without cloud support	Phones, smart cameras

The Chip Family reminds us that behind every intelligent system is powerful hardware performing millions or billions of calculations, turning abstract models into real-world functionality.

Watts Family Introduction

The Watts Family represents the measurement of power used by artificial intelligence systems. While the Chip Family provides the hardware that performs computations, the Watts Family explains how much energy those systems consume while operating.

A watt is a unit of power that measures the rate at which energy is used. In artificial intelligence, watts help us understand how much electricity is required to run processors, train models, and generate responses. Every time an AI system processes data, performs calculations, or produces output, it consumes power.

Different AI systems require different levels of power depending on their size and complexity. Small systems, such as those running on mobile devices, use relatively low amounts of power. In contrast, large-scale AI models operating in data centers can require significant amounts of electricity, especially during training, where billions of calculations are performed continuously.

The Watts Family works closely with the Chip Family, which determines how efficiently computations are performed, and the Infrastructure Family, where large networks of machines operate together. It also connects to the overall design of AI systems, as more efficient models can reduce power consumption while maintaining performance.

Understanding the Watts Family helps students recognize that artificial intelligence is not only about intelligence and computation, but also about energy use and resource management. As AI continues to grow, managing power efficiently becomes an important part of building responsible and sustainable systems.

In simple terms, if chips perform the work, watts measure how much power it takes to do that work.

Watts Family Breakdown Chart

Component	Role	Description	Example
Watt (W)	Power Measurement	Unit that measures the rate of energy use	A device using 100 watts
Power Consumption	Energy Usage Rate	Amount of power used during operation	AI model running on GPUs
Energy Efficiency	Performance Balance	How effectively a system uses power to perform tasks	More output with less power
High-Performance Computing	Intensive Power Use	Systems that require large amounts of power for complex tasks	AI training clusters
Idle Power	Standby Usage	Power consumed when systems are not actively processing	Servers waiting for requests
Peak Power	Maximum Usage	Highest level of power used during heavy workloads	Training large AI models
Thermal Output	Heat Generation	Heat produced as a result of power usage	Cooling systems in data centers
Cooling Systems	Temperature Control	Systems used to manage heat from high power usage	Air or liquid cooling
Power Supply	Energy Source	Provides electricity to AI systems	Electrical grid, batteries
Edge Power Usage	Local Energy Use	Power consumption on smaller devices	Smartphones, IoT devices
Data Center Power	Large-Scale Usage	Power required for large AI operations	Server farms running AI models

Component	Role	Description	Example
Sustainable Energy	Efficiency Goal	Use of renewable energy to reduce environmental impact	Solar-powered data centers

The Watts Family highlights that intelligence comes with a cost—every AI system requires power, and understanding that power is essential for building efficient and sustainable technologies.

Watts Family (AI Power Usage Scale)

Definition

The **Watts Family** represents the real-time power usage of artificial intelligence systems, measured in watts (W). It shows how much power AI systems consume at a given moment as they process data and perform computations.

Watts Scale

1 Watt
10 Watts
100 Watts
500 Watts
1,000 Watts (1 Kilowatt)
10,000 Watts (10 Kilowatts)
100,000 Watts (100 Kilowatts)
1,000,000 Watts (1 Megawatt)
10,000,000 Watts (10 Megawatts)
100,000,000+ Watts (100+ Megawatts)

Key Characteristics

Measured in real-world electrical power

Higher watt values indicate higher processing demand

Used to represent the scale of AI systems from small devices to large infrastructures

Closely connected to the Chip Family, Energy Family, and Infrastructure Family

Examples

Low watts → small devices and basic AI functions

Medium watts → computers and advanced applications

High watts → data centers and large AI systems

Summary

The **Watts Family** provides a numerical representation of how much power artificial intelligence systems use in real time, helping to illustrate the scale and intensity of AI operations.

🧠 Student Insight

Watts show how much power AI is using right now.

AI Energy Family Introduction

The AI Energy Family represents the total amount of energy consumed by artificial intelligence systems over time. While the Watts Family measures the rate at which power is used at any given moment, the Energy Family focuses on the accumulated energy required to perform tasks, run systems, and sustain operations.

In artificial intelligence, energy is used every time a model is trained, a system processes data, or a response is generated. Large-scale AI systems, especially those operating in data centers, can consume significant amounts of energy due to the continuous processing of massive datasets and complex computations.

Energy is typically measured over time using units such as kilowatt-hours (kWh), which reflect how much power has been used during a specific period. For example, running a high-performance AI system for several hours or days results in a total energy cost that goes beyond the moment-to-moment power usage measured in watts.

The AI Energy Family works closely with the Watts Family, which provides the rate of power consumption, and the Chip Family, which determines how efficiently computations are performed. It also connects to the Infrastructure Family, where large-scale systems operate continuously and require long-term energy management.

Understanding the AI Energy Family helps students recognize that artificial intelligence has real-world resource implications. It is not only about performance and speed, but also about sustainability, efficiency, and responsible use of technology.

In simple terms, if watts measure how fast energy is used, energy measures how much is used over time.

AI Energy Family Breakdown Chart

Component	Role	Description	Example
Energy	Total Consumption	The total amount of power used over a period of time	Running an AI system for hours or days
Kilowatt-hour (kWh)	Energy Measurement	Unit used to measure energy consumption over time	1 kWh = using 1,000 watts for 1 hour
Energy Usage	Consumption Tracking	Total energy required for AI operations	Training a large model
Training Energy	Learning Cost	Energy used during model training processes	Weeks of GPU usage
Inference Energy	Response Cost	Energy used when AI generates outputs	Answering a user prompt
Energy Efficiency	Optimization Goal	Reducing energy use while maintaining performance	Efficient AI models
Energy Scaling	Growth Impact	Increase in energy use as systems grow larger	Bigger models = more energy
Data Center Energy	Infrastructure Demand	Total energy used by large AI facilities	Server farms operating 24/7
Cooling Energy	Temperature Control	Energy required to cool systems and prevent overheating	Air conditioning in data centers
Renewable Energy	Sustainability Source	Use of clean energy to power AI systems	Solar, wind-powered data centers
Carbon Impact	Environmental Effect	Emissions associated with energy consumption	AI training carbon footprint
Energy Management	Resource Control	Strategies to monitor and reduce energy use	Efficient scheduling, hardware optimization

The AI Energy Family reminds us that artificial intelligence operates within the physical world, where every computation consumes energy and every system has an impact on resources and sustainability.

Energy Family (Total Power Consumption of Artificial Intelligence)

Definition

The **Energy Family** represents the total amount of electricity used by artificial intelligence systems over time. It is measured in watt-hours (Wh) or kilowatt-hours (kWh) and shows how much energy AI systems consume while running.

Energy Scale

1 Watt-hour (Wh)
10 Watt-hours (Wh)
100 Watt-hours (Wh)
500 Watt-hours (Wh)
1,000 Watt-hours (1 Kilowatt-hour, kWh)
10,000 Watt-hours (10 kWh)
100,000 Watt-hours (100 kWh)
1,000,000 Watt-hours (1 Megawatt-hour, MWh)
10,000,000 Watt-hours (10 MWh)
100,000,000+ Watt-hours (100+ MWh)

Key Characteristics

Measured over time (not instant like watts)

Represents total electricity consumption

Increases the longer AI systems run

Directly related to cost and energy usage

Closely connected to Watts Family, Chip Family, and Infrastructure Family

Examples

Low energy → short AI tasks or small devices

Medium energy → daily AI usage on computers

High energy → training AI models and running data centers

Summary

The **Energy Family** provides a numerical representation of the total electricity consumed by artificial intelligence systems over time, helping to illustrate the overall cost and impact of AI operations.

🐾 Student Insight

Energy shows how much total power AI has used over time.

AI Infrastructure Family Introduction

The AI Infrastructure Family represents the large-scale systems and environments that support, power, and connect artificial intelligence technologies. While individual components such as chips, data, and models explain how AI works, infrastructure explains where and how these systems operate in the real world.

AI infrastructure includes data centers, cloud platforms, networking systems, and storage environments that allow artificial intelligence to function at scale. These systems provide the computing power, data access, and connectivity needed to train models, process information, and deliver results to users around the world.

Modern AI systems rely heavily on cloud-based infrastructure, where thousands of machines work together to handle massive workloads. This allows AI to scale efficiently, making it possible to process large datasets, run complex models, and serve millions of users simultaneously. Infrastructure also ensures reliability, security, and continuous availability of AI services.

The AI Infrastructure Family works closely with the Chip Family, which provides the hardware, the Watts and Energy Families, which manage power consumption, and the Data Family, which supplies the information being processed. It also supports the AI Architecture Family by providing the environment where all system components are deployed and connected.

Understanding the AI Infrastructure Family helps students recognize that artificial intelligence is not contained within a single device. Instead, it operates across global systems that require coordination, resources, and engineering to function effectively.

In simple terms, if AI is the system and chips are the engines, infrastructure is the environment that allows everything to run at scale.

AI Infrastructure Family Breakdown Chart

Component	Role	Description	Example
AI Infrastructure	System Environment	The physical and digital systems that support AI operations	Global AI platforms
Data Center	Processing Hub	Facilities that house servers and computing equipment	Large server farms
Cloud Computing	Scalable Platform	Remote systems that provide computing resources over the internet	AWS, Azure, Google Cloud
Servers	Compute Units	Machines that process data and run AI models	Rack-mounted servers

Component	Role	Description	Example
Networking	Connectivity System	Systems that allow communication between machines	Internet, fiber networks
Storage Systems	Data Management	Systems that store large volumes of data	Databases, cloud storage
Distributed Computing	Workload Sharing	Splitting tasks across multiple machines for efficiency	Parallel processing across clusters
Edge Infrastructure	Local Processing	Running AI closer to the user or device	Smart devices, IoT systems
Load Balancing	Traffic Control	Distributes workloads to prevent overload	Managing user requests
Security Systems	Protection Layer	Safeguards data and systems from threats	Encryption, firewalls
Redundancy	Reliability System	Backup systems to ensure continuous operation	Failover servers
Scalability	Growth Capability	Ability to expand resources as demand increases	Adding more servers dynamically

The AI Infrastructure Family reveals that artificial intelligence operates on a global scale, relying on interconnected systems that provide the power, storage, and connectivity needed to support intelligent technologies.

AI Agent Family Introduction

The AI Agent Family represents systems that can take action, make decisions, and perform tasks on behalf of users or other systems. Unlike traditional AI models that only respond to input, AI agents are designed to operate with a level of autonomy, allowing them to plan, execute, and adapt to achieve specific goals.

An AI agent receives input from its environment, processes that information, and takes actions based on predefined objectives or learned behaviors. These actions can include answering questions, automating workflows, retrieving information, or interacting with other systems. Some agents operate in simple environments, while others function in complex systems that require continuous decision-making.

AI agents often combine multiple foundational families. They use the Data Family to gather information, the Memory and Knowledge Families to retain and understand context, and the Neural Network and Parameter Families to process and learn from interactions. They also rely on the AI Architecture and Infrastructure Families to operate reliably at scale.

Modern AI agents can work independently or as part of larger systems. For example, a customer support agent can respond to inquiries automatically, while a more advanced agent can complete multi-step tasks such as scheduling, research, or system management.

Understanding the AI Agent Family helps students see that artificial intelligence is not only about generating responses, but also about taking meaningful action in real-world scenarios.

In simple terms, if AI models think and respond, AI agents act and execute.

"Learn AI. Understand AI. Shape the Future."
— Susie Hala

AI Agent Family Breakdown Chart

Component	Role	Description	Example
AI Agent	Action System	An AI system that can make decisions and perform tasks	Virtual assistant
Environment	Operating Context	The space where the agent interacts and gathers information	Web, apps, databases
Input Perception	Data Intake	Receives information from the environment	User query or sensor input
Decision Engine	Action Selection	Determines what action to take based on input	Choosing a response or task
Action Execution	Task Performance	Carries out decisions in the environment	Sending an email, retrieving data
Goal System	Objective Setting	Defines what the agent is trying to achieve	Complete a task or solve a problem
Planning Module	Strategy Builder	Breaks down tasks into steps	Multi-step problem solving
Feedback Loop	Learning Cycle	Improves performance based on outcomes	Adjusting actions over time
Autonomy Level	Independence Scale	Degree of independence in decision-making	Fully automated vs assisted
Multi-Agent System	Collaboration Network	Multiple agents working together	Coordinated AI systems

Ingredient Family Introduction

The Ingredient Family represents the essential components that come together to create an artificial intelligence system. Just as a recipe requires specific ingredients to produce a final dish,

AI systems rely on a combination of foundational elements working together to function effectively.

Artificial intelligence is not built from a single component. It requires data to learn from, algorithms to guide processing, models to perform tasks, hardware to execute computations, and energy to power the entire system. Each of these elements plays a critical role, and the absence of any one component would prevent the system from operating properly.

The Ingredient Family provides a simplified way for students to understand AI as a system made up of interconnected parts. Instead of viewing AI as complex or mysterious, learners can begin to see it as something constructed from identifiable and understandable components.

This family connects directly with all other foundational families, including Data, Algorithm, Neural Network, Parameter, Chip, and Energy Families. It serves as an overview that brings these elements together into a unified perspective.

Understanding the Ingredient Family helps students recognize that artificial intelligence is built, not magical. It is the result of carefully combining the right components in the right way.

In simple terms, if AI is the final product, the Ingredient Family represents everything needed to build it.

Ingredient Family Breakdown Chart

Component	Role	Description	Example
Data	Learning Input	Provides the information AI uses to learn patterns	Text, images, audio
Algorithms	Instruction Set	Guides how data is processed and decisions are made	Sorting, classification
AI Models	Execution System	Performs tasks such as prediction or generation	Language models
Neural Networks	Processing Structure	Organizes how data is processed through layers	Deep learning models
Parameters	Learning Adjustments	Fine-tune how the model learns and makes predictions	Weights and biases
Tokens	Language Units	Breaks text into smaller pieces for processing	Words or subwords
Hardware (Chips)	Computation Engine	Executes calculations required by AI systems	GPUs, CPUs
Energy	Power Source	Supplies the energy needed to run systems	Electricity usage
Infrastructure	Support System	Provides the environment for AI to operate at scale	Cloud systems, data centers

Component	Role	Description	Example
Memory	Storage System	Stores and retrieves information for processing	Context memory
Knowledge	Understanding Layer	Represents meaning and relationships within data	Knowledge graphs

The Ingredient Family shows that artificial intelligence is built from a combination of essential components, each playing a vital role in creating intelligent systems.

"Learn AI. Understand AI. Shape the Future."
— Susie Hala

The Frontend and Backend Family of Artificial Intelligence

How AI Systems Connect User Interaction to Intelligent Processing

Introduction

Every artificial intelligence system operates through two essential layers: the frontend and the backend. These layers work together to transform human interaction into intelligent responses, allowing AI systems to function smoothly and effectively.

The frontend is the part of the system that users see and interact with. It includes screens, input tools, and the display of results. Whether a user is typing a question, speaking a command, or uploading an image, all interaction begins at the frontend.

Behind the scenes, the backend serves as the processing engine of the system. It receives input from the frontend, manages data, connects to the AI model, and performs the computations needed to generate a response. The backend is where the intelligence of the system operates, using models, algorithms, and infrastructure to interpret and respond to user requests.

This visual illustrates how these two layers work together as a complete system. It shows the flow of information from user input to AI processing and back to the user as a result. By understanding this structure, students can see that AI is not a single tool, but a coordinated system of layers working together.

In simple terms, the frontend is where interaction happens, and the backend is where intelligence is created.

Every AI system has two sides: what you see and what actually does the thinking.

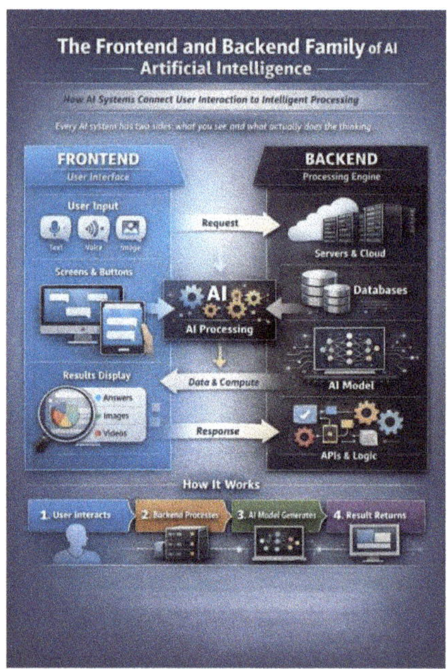

AI Qubit Family Introduction

The AI Qubit Family represents the foundation of quantum computing as it relates to artificial intelligence. While traditional computing relies on bits, which can exist as either 0 or 1, quantum computing uses qubits, which can exist in multiple states simultaneously.

A qubit is the basic unit of quantum information. Unlike classical bits, qubits can take advantage of properties such as superposition and entanglement, allowing quantum systems to process complex computations more efficiently in certain scenarios. This opens new possibilities for solving problems that are difficult or impossible for traditional computers.

In the context of artificial intelligence, qubits have the potential to accelerate learning processes, optimize large-scale systems, and improve complex simulations. Although quantum AI is still in development, it represents a future direction where computational power can expand beyond current limitations.

The AI Qubit Family connects with the Chip Family, as quantum processors are specialized hardware, and with the Algorithm Family, where new types of quantum algorithms are designed. It also relates to the Parameter and Neural Network Families, as researchers explore quantum-enhanced learning models.

Understanding the AI Qubit Family helps students see that artificial intelligence is continuously evolving. It introduces the idea that the foundations of computing themselves can change, leading to new forms of intelligence and problem-solving.

In simple terms, if bits are the foundation of today's computing, qubits represent the foundation of tomorrow's possibilities.

AI Qubit Family Breakdown Chart

Component	Role	Description	Example
Qubit	Quantum Unit	Basic unit of quantum information	Quantum bit in a quantum computer
Superposition	Multi-State Ability	Ability to exist in multiple states at once	0 and 1 simultaneously
Entanglement	Connection Property	Linking qubits so they influence each other	Paired quantum states
Quantum Gate	Operation System	Performs operations on qubits	Quantum logic gates
Quantum Circuit	Processing Structure	Sequence of quantum operations	Quantum algorithms
Quantum Processor	Hardware Engine	Specialized chip for quantum computing	Quantum computer hardware
Quantum Algorithm	Computation Method	Algorithm designed for quantum systems	Optimization problems
Quantum Speedup	Performance Advantage	Faster processing for certain problems	Complex simulations
Quantum Noise	Stability Challenge	Errors caused by environmental interference	Qubit instability
Quantum Error Correction	Stability Solution	Techniques to reduce errors in quantum systems	Fault-tolerant computing

The AI Qubit Family introduces a new frontier in computing, where the limits of traditional systems are expanded, opening the door to more powerful and advanced forms of artificial intelligence.

"Learn AI. Understand AI. Shape the Future."
— Susie Hala

AI Tools Used in K-12 Classrooms

Empowering Students and Educators with Innovative AI Tools for Teaching and Learning

 **Adobe Express**

Create graphics, videos, presentations, and more with AI-powered design tools.

 **Book Creator**

Create digital books and stories with the help of AI-assisted content tools.

 Canva

Design presentations, posters, and visual content with AI suggestions.

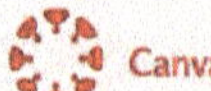 Canvas

AI-enhanced learning management system to support instruction and engagement.

 Copilot

AI assistant by Microsoft that helps with research, writing, and problem solving.

 Gemini

AI from Google that supports learning, research, writing, and creativity.

 Nearpod

AI-powered lessons, interactive activities, and real-time student engagement.

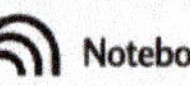 NotebookLM

AI-research assistant that helps summarize information and generate insights.

 OneNote

AI features that help organize notes, ideas, and classroom collaboration.

 Padlet

AI-enhanced digital boards for collaboration, brainstorming, and sharing ideas.

 PowerPoint

AI-powered design ideas, smart layouts, and presentation enhancements.

 Reading Coach

AI tool that helps students build reading fluency and comprehension.

 Whiteboard

AI-powered digital whiteboard for collaboration and idea generation.

 **Microsoft Forms**

AI-assisted quizzes, surveys, and feedback collection.

 **Microsoft Teams**

AI features like transcriptions, summaries, and smart meeting notes.

 Sway

AI-powered digital storytelling and interactive presentations.

 **Immersive Reader**

AI tool that supports reading comprehension and accessibility for all learners.

 Flip

AI-powered video discussions and engagement for the classroom.

 **Note:** These AI tools are examples of platforms used to enhance teaching, support learning, and promote creativity. Educators should review each tool for student privacy, safety, and school or district approval before classroom use.

 **IntelliGloss**

The AI-Ready Classroom

Preparing Schools and Educators to Teach Artificial Intelligence Responsibly

INTRODUCTION

Teaching in a Time of Intelligent Machines

Artificial intelligence is no longer a distant concept confined to research laboratories or science fiction. It has entered classrooms, homework assignments, school policies, and everyday student life. From AI-powered search tools to generative writing assistants, students are already interacting with intelligent systems every day—often without fully understanding how those systems work.

For educators, this rapid shift raises important questions:

• How should AI be addressed in the classroom?
• Should it be restricted, integrated, or redesigned around?
• How can teachers preserve academic integrity while preparing students for an AI-driven future?

These questions are not signs of uncertainty; they are signs of responsible leadership.

The purpose of this book is not to promote artificial intelligence uncritically, nor to dismiss legitimate concerns. Instead, this guide offers a structured framework for understanding, evaluating, and implementing AI literacy in grades 6–12 education.

Artificial intelligence is not simply a new classroom tool. It is a transformative force reshaping communication, research, creativity, and problem-solving. Just as educators once guided students through the transition from print to digital learning, today's teachers are uniquely positioned to guide students through the transition from digital tools to intelligent systems.

This transition requires thoughtful leadership. Schools must balance innovation with responsibility, curiosity with caution, and opportunity with ethical awareness.

The **AI-Ready Classroom** is built on three core principles:

Clarity before adoption

Responsibility before reliance

Critical thinking before automation

These principles ensure that artificial intelligence becomes a tool for deeper learning rather than a shortcut that weakens intellectual growth.

This book provides:

• A clear explanation of how artificial intelligence works
• A practical classroom framework for AI literacy
• A 16-week sample implementation model
• Assessment tools and rubrics
• Sample policy language for schools and districts
• Ethical considerations for responsible AI use in education

AI literacy is not about turning every student into a programmer. It is about ensuring students understand how intelligent systems function, where their limitations exist, and how to use them responsibly and thoughtfully.

Educators do not need to become engineers to teach AI literacy. They need structure, clarity, and confidence. This guide is designed to provide all three.

Artificial intelligence will continue evolving. New tools will emerge, and educational policies will adapt. Yet the central mission of education remains unchanged: to cultivate thoughtful, capable, and ethical individuals who can navigate a complex world.

The future of education will not be defined by whether AI enters the classroom. It already has.

The defining question is how educators choose to lead within it.

"Learn AI. Understand AI. Shape the Future."
— Susie Hala, IntelliGloss Press

Chapter 1

Common Teacher Concerns in the Age of Artificial Intelligence

Artificial intelligence did not gradually enter the classroom through carefully planned curriculum revisions. It arrived quickly—through student smartphones, homework submissions, and online tools that could generate essays, solve equations, and summarize research in seconds. For many educators, this shift felt abrupt and destabilizing.

Before discussing integration strategies or curriculum frameworks, it is essential to acknowledge the very real concerns teachers are experiencing. These concerns are not obstacles to progress; they are thoughtful responses to rapid change.

This chapter addresses the most common questions educators raise and provides clarity that supports informed decision-making.

Concern 1: "Is AI Just a New Way for Students to Cheat?"

Academic integrity is one of the first issues that arises when generative AI tools become accessible. If a system can produce an essay or solve a problem instantly, how can teachers assess authentic student learning?

This concern is understandable. However, artificial intelligence did not create academic dishonesty. It changed the method by which it can occur. Similar fears arose when calculators, internet search engines, and online translation tools first entered schools.

The deeper question is not whether AI can be misused. It can. The more productive question is how assessment models must evolve in response.

Rather than relying solely on traditional take-home written assignments, educators can incorporate:

- In-class drafting and reflection
- Oral defense of written work
- Project-based learning
- Process documentation
- Iterative revisions
- AI-use disclosure statements

Artificial intelligence challenges educators to evaluate not only the final product, but also the thinking process behind it. When assessment shifts toward reasoning, explanation, and creativity, misuse becomes more difficult and authentic learning becomes more visible.

Concern 2: "Will Students Stop Thinking for Themselves?"

Another widespread fear is cognitive dependency. If students can ask AI to summarize texts, generate ideas, or solve equations, will they lose critical thinking skills?

This concern touches on an important principle: tools can either enhance thinking or replace it, depending on how they are used.

When AI is treated as a shortcut, thinking decreases.
When AI is treated as a thinking partner, analysis deepens.

For example, students can be asked to:

- Compare their own response with an AI-generated response
- Critique AI output for bias or errors
- Improve AI responses using better prompts
- Identify limitations in AI reasoning

These activities strengthen metacognition. Instead of passively accepting information, students evaluate it.

Artificial intelligence does not eliminate the need for thinking. It increases the need for discernment.

Concern 3: "Will AI Replace Teachers?"

Technological change often generates professional anxiety. When intelligent systems can explain concepts, generate lesson plans, or provide tutoring support, some educators wonder whether their role will diminish.

History suggests otherwise.

When calculators were introduced, math teachers did not disappear. Instead, instruction shifted toward higher-level problem-solving. When the internet became widespread, teachers did not become obsolete; they became guides in evaluating information credibility.

Artificial intelligence may automate certain tasks, such as drafting outlines or generating practice questions. However, it cannot replace:

- Emotional intelligence
- Relationship-building
- Classroom leadership
- Ethical guidance
- Cultural responsiveness
- Adaptive instruction based on human nuance

The teacher's role evolves, but it does not vanish. In fact, in an AI-rich environment, human guidance becomes more essential, not less.

Concern 4: "Is Student Data Safe?"

Privacy concerns are legitimate. Many AI systems require user input and may collect data for model improvement.

Schools must approach AI adoption with careful attention to:

- Data protection policies
- Student privacy laws
- Platform compliance with educational standards
- Clear district-level guidelines

AI literacy does not require schools to adopt every available tool. It requires informed evaluation of which tools align with institutional safeguards.

Responsible implementation begins with policy, not experimentation.

Concern 5: "I Was Never Trained for This."

Perhaps the most honest concern teachers express is uncertainty. Many educators did not receive formal training in artificial intelligence during their credential programs. The pace of technological advancement can feel overwhelming.

However, AI literacy does not require technical engineering expertise. It requires conceptual clarity.

Teachers do not need to code machine learning models. They need to understand:

- What AI can and cannot do
- How AI generates responses
- Where AI makes errors
- How to guide responsible student use

Professional growth in AI education can begin with foundational understanding, not advanced programming.

Reframing the Conversation

Each of these concerns reflects responsible professionalism. Artificial intelligence introduces complexity into education, but complexity is not a reason for avoidance. It is a reason for structured response.

Avoidance leaves students to navigate AI independently.
Prohibition without education creates underground usage. Guided
literacy fosters responsible engagement.

The purpose of this book is not to dismiss teacher concerns, but to build confidence through clarity.

Artificial intelligence will continue evolving. The question is not whether schools can prevent exposure. The question is whether educators will shape how students understand and use it.

The AI-ready classroom begins not with adoption, but with informed leadership.

"Learn AI. Understand AI. Shape the Future."
— Susie Hala, IntelliGloss Press

Chapter 2

What Artificial Intelligence Actually Is

Artificial intelligence is often discussed in headlines, social media debates, and professional development sessions. Yet despite frequent discussion, the term itself is frequently misunderstood. Before educators can design policy, adapt instruction, or guide students responsibly, it is essential to establish a clear understanding of what artificial intelligence is—and what it is not.

Artificial intelligence refers to computer systems designed to perform tasks that typically require human intelligence. These tasks may include recognizing patterns, generating language, identifying images, making predictions, or recommending actions based on data.

AI does not think in the human sense. It does not possess consciousness, emotions, or independent intent. Instead, it processes data using mathematical models that identify patterns and generate outputs based on probability.

Understanding this distinction is foundational to AI literacy.

Narrow AI: The Systems We Use Today

Most artificial intelligence tools currently used in education fall under the category of Artificial Narrow Intelligence (ANI).

Narrow AI is designed to perform specific tasks. It excels within defined boundaries but does not generalize beyond its programming.

Examples include:

- Language generation tools that produce written text
- Recommendation systems that suggest content
- Image recognition systems that identify objects
- Adaptive learning platforms that adjust difficulty levels

These systems do not understand content in a human sense. They identify patterns in large datasets and generate outputs that statistically align with user input.

When a generative AI tool produces an essay, it is not expressing opinion or original thought. It is predicting the most likely sequence of words based on training data patterns.

This probabilistic nature explains both the strengths and the limitations of AI systems.

How Generative AI Produces Responses

Generative AI models are trained on large datasets consisting of text, images, code, or other digital information. Through repeated exposure to patterns within that data, the system adjusts internal mathematical parameters to improve its predictions.

When a user enters a prompt, the model does not "search the internet" in real time. Instead, it generates a response by predicting what sequence of words is most statistically likely to follow the prompt, based on patterns learned during training.

This means:

- AI responses may appear fluent and confident.
- AI responses may contain inaccuracies.
- AI systems do not verify truth unless connected to external tools.

The appearance of intelligence does not equal understanding.

This is a crucial distinction for educators and students alike.

Machine Learning: The Engine Behind AI

Artificial intelligence systems often rely on machine learning, a method in which algorithms improve performance through exposure to data.

Rather than being programmed with explicit instructions for every scenario, machine learning models adjust internal parameters based on examples.

For example:

- A model trained to recognize handwritten digits analyzes thousands of labeled examples.
- Over time, it adjusts mathematical weights to improve accuracy.

This process allows AI systems to improve within specific domains. However, the improvement remains bounded by the scope of training data and model design.

Machine learning does not create independent reasoning. It refines pattern recognition.

Artificial General Intelligence: A Theoretical Concept

Public discussion sometimes references Artificial General Intelligence (AGI), which would represent a system capable of performing any intellectual task that a human can perform.

At present, AGI does not exist.

All AI systems currently deployed in classrooms, businesses, and homes operate within narrow task boundaries. While some tools appear versatile, they remain pattern-based systems without self-awareness or genuine comprehension.

Distinguishing between existing narrow AI and theoretical general AI helps prevent exaggerated expectations and unnecessary alarm.

Strengths and Limitations of AI Systems

Understanding AI requires acknowledging both capabilities and constraints.

AI systems are strong at:

- Processing large volumes of data
- Identifying patterns
- Generating structured responses
- Performing repetitive analytical tasks

AI systems are limited by:

- Lack of true understanding
- Potential bias in training data
- Inability to verify facts independently
- Overconfidence in inaccurate outputs
- Absence of ethical reasoning

Students must learn not only how to use AI tools, but how to evaluate them critically.

Why Conceptual Clarity Matters in Education

Without clear definitions, artificial intelligence becomes either mystified or feared. Overestimation leads to dependency. Underestimation leads to misuse.

AI literacy begins with accurate understanding:

- AI generates predictions, not thoughts.
- AI simulates conversation, not consciousness.
- AI assists with tasks but does not replace human judgment.

When educators understand the mechanisms behind AI tools, they are better equipped to design assignments, develop policies, and model responsible use.

Clarity reduces anxiety.
Clarity supports leadership.

Artificial intelligence is not magic, and it is not menace. It is a technological system built on data, algorithms, and statistical prediction.

In the next chapter, we move from definition to application: why AI literacy has become essential in grades 6–12 education and how schools can respond proactively rather than reactively.

"Learn AI. Understand AI. Shape the Future."
— Susie Hala, IntelliGloss Press

Chapter 3

Why AI Literacy Matters in K–12 Education

Artificial intelligence is not a future concept awaiting classroom introduction. It is already embedded in the tools students use daily. Search engines suggest answers before questions are finished. Writing assistants draft paragraphs in seconds. Recommendation systems influence what students read, watch, and purchase. Adaptive learning platforms adjust difficulty based on performance.

Whether schools formally teach artificial intelligence or not, students are interacting with it.

The question facing education is not whether students will encounter AI. It is whether they will understand it.

AI literacy is not about turning students into engineers. It is about equipping them with the awareness, critical thinking, and ethical reasoning necessary to navigate intelligent systems responsibly.

Students Are Already Using AI

Many students experiment with AI tools outside of school before educators even discuss them in class. When access exists without guidance, several outcomes may occur:

- Overreliance on automated outputs
- Acceptance of information without verification
- Misunderstanding of AI capabilities
- Academic misuse
- Exposure to biased or inaccurate responses

Avoidance does not prevent usage. Silence does not prevent experimentation.

AI literacy creates a structured space for discussion, evaluation, and responsible engagement.

AI Literacy Is a Form of Digital Citizenship

For decades, schools have taught digital citizenship—how to evaluate sources, protect privacy, and engage respectfully online. Artificial intelligence extends this responsibility.

Students must now learn:

- How AI generates responses
- Why AI may produce errors
- How bias can appear in automated systems
- When AI assistance is appropriate
- How to disclose AI use transparently

AI literacy builds discernment. It strengthens the ability to question rather than consume.

When students understand how intelligent systems function, they are less likely to treat outputs as unquestionable authority.

Preparing Students for an AI-Influenced Workforce

Workplace transformation is accelerating. Artificial intelligence tools are increasingly integrated into fields such as healthcare, finance, engineering, marketing, law, and creative industries.

Employers are not simply seeking individuals who can use AI tools. They are seeking individuals who can:

- Interpret AI-generated results
- Identify limitations
- Apply human judgment
- Collaborate with automated systems
- Solve problems AI cannot address independently

AI literacy develops these higher-order competencies.

Schools that integrate structured AI understanding are not replacing traditional subjects. They are strengthening cross-disciplinary skills that apply across subjects and careers.

Strengthening Critical Thinking Through AI

Contrary to the assumption that AI weakens cognition, structured AI engagement can enhance critical thinking.

Students can be asked to:

- Compare their original writing to AI-generated writing

- Identify inaccuracies in AI responses
- Revise AI outputs for clarity or correctness
- Debate ethical implications of AI decisions
- Analyze how prompt wording influences results

These exercises shift students from passive consumers to analytical evaluators.

Artificial intelligence becomes a case study in reasoning, not a substitute for it.

Equity and Access Considerations

Access to artificial intelligence tools is uneven. Some students encounter advanced systems regularly, while others may lack exposure. Without structured instruction, this disparity can widen.

AI literacy ensures that:

- All students receive foundational understanding
- Students learn safe and responsible usage practices
- Critical thinking skills are developed regardless of background
- Opportunities for informed engagement are equitable

Avoiding AI in education does not eliminate inequality. Structured instruction can help address it.

Moving From Reaction to Leadership

Educational institutions have historically adapted to technological change. Calculators altered mathematics instruction. The internet transformed research practices. Digital collaboration reshaped communication.

Artificial intelligence represents another inflection point.

Schools can respond in one of three ways:

1. Prohibition without instruction
2. Unregulated adoption
3. Structured literacy and guided integration

The third approach positions educators as leaders rather than reactors.

AI literacy does not mean endorsing every tool. It means equipping students to evaluate tools thoughtfully and ethically.

The Role of Educators

Teachers are not expected to become AI engineers. They are expected to guide learning.

Guidance begins with understanding. When educators possess conceptual clarity, they can model:

- Informed skepticism
- Responsible experimentation

Ethical awareness
    Intellectual integrity

Artificial intelligence will continue evolving. The pace of development may challenge traditional curriculum cycles. However, the core educational mission remains constant: to cultivate thoughtful, capable, and ethical individuals.

AI literacy is an extension of that mission.

In the next chapter, we introduce a structured framework for building an AI-ready classroom— one that balances clarity, responsibility, and critical thinking

"Learn AI. Understand AI. Shape the Future."
— Susie Hala, IntelliGloss Press

Chapter 4

The IntelliGloss AI Literacy Framework

Artificial intelligence literacy cannot rely on isolated lessons or reactive policies. It requires a structured approach that integrates understanding, responsibility, and critical engagement into classroom practice.

The IntelliGloss AI Literacy Framework is designed to provide that structure.

This framework does not center on specific platforms or software tools. Instead, it focuses on durable competencies that remain relevant even as technology evolves. By grounding instruction in foundational principles rather than temporary applications, schools can build sustainable AI literacy programs.

The IntelliGloss framework is built on four pillars:

1. Conceptual Understanding
2. Responsible Use
3. Critical Evaluation
4. Creative & Collaborative Application

Together, these pillars form a balanced model for grades 6–12 instruction.

Pillar One: Conceptual Understanding

Before students can use artificial intelligence responsibly, they must understand how it works.

Conceptual understanding includes:

- Recognizing the difference between narrow AI and theoretical general AI

    Understanding that AI generates outputs through pattern recognition
    Knowing that AI systems can produce errors
- Recognizing the influence of training data on outcomes

Students should learn that AI does not "know" information in a human sense. It predicts responses based on probability. This foundational clarity reduces overreliance and encourages thoughtful engagement.

At the middle school level, conceptual instruction may focus on simplified explanations of pattern recognition and prediction.

*
*

At the high school level, instruction may expand to include discussions of algorithms, machine learning models, bias, and data limitations.

Conceptual understanding is the anchor that prevents misunderstanding.

Pillar Two: Responsible Use

Artificial intelligence introduces ethical considerations into everyday learning.

Responsible use includes:

- Transparency in AI assistance
- Proper disclosure when AI tools contribute to work
- Respect for academic integrity
- Awareness of privacy concerns
- Understanding school and district policies

Students must learn when AI support is appropriate and when independent work is required.

Teachers can model responsible use by:

- Demonstrating how AI can assist brainstorming without replacing original thinking
- Discussing appropriate citation practices
- Establishing clear classroom AI guidelines

Responsible use reinforces that AI is a tool, not a replacement for personal accountability.

Pillar Three: Critical Evaluation

Artificial intelligence systems can generate confident but inaccurate responses. Therefore, critical evaluation becomes essential.

Students should be taught to:

- Verify AI-generated information

  Compare AI responses with credible sources
  Identify bias or incomplete perspectives
- Recognize hallucinations or fabricated references
- Evaluate the clarity and logic of outputs

Rather than discouraging AI interaction, educators can use it as a teaching opportunity.

•
•

For example:

Students might analyze an AI-generated essay and identify weaknesses in argument structure or evidence. This shifts students from passive recipients to analytical reviewers.

Critical evaluation strengthens reasoning skills across disciplines.

Pillar Four: Creative & Collaborative Application

Artificial intelligence can serve as a creative partner when used appropriately.

In structured settings, students may:

- Use AI to generate multiple perspectives on a topic
- Refine ideas through iterative prompting
- Develop project outlines
- Explore alternative problem-solving approaches
- Simulate debate scenarios

The goal is not to outsource thinking but to enhance it.

Collaborative AI use requires students to reflect on how the tool influenced their work. Reflection questions might include:

- How did AI support your thinking?
- What limitations did you observe?
- How did you revise AI-generated content to reflect your own voice?

When creativity is paired with reflection, AI becomes an instructional catalyst rather than a shortcut.

Integrating the Four Pillars

The strength of the IntelliGloss AI Literacy Framework lies in balance.

Conceptual understanding without responsibility risks misuse.
Responsible use without evaluation risks blind trust.
Evaluation without creativity limits innovation.
Creativity without structure risks dependency.

Together, the four pillars create a comprehensive model that prepares students not just to use AI tools, but to understand and manage them thoughtfully.

Grade-Level Adaptation

The framework is designed to scale across grade levels.

Middle School Focus:

- Basic AI definitions
- Introduction to responsible use • Guided evaluation exercises
- Structured AI-assisted projects

High School Focus:

- Deeper exploration of machine learning concepts
- Ethical debates
- Advanced evaluation of bias and misinformation
- Complex project-based collaboration

This progression supports developmental readiness while maintaining consistent structure.

A Sustainable Model

Technology will continue evolving. Specific platforms may rise and fall. However, the competencies outlined in this framework remain stable:

- Understanding
- Responsibility
- Evaluation
- Creation

Schools that adopt a principle-based model rather than a tool-based model position themselves for long-term adaptability.

The IntelliGloss AI Literacy Framework is not a temporary response to a trend. It is a structured educational approach designed to evolve alongside technological change.

In the next chapter, we translate this framework into a practical implementation plan, including scope, sequence, and a sample instructional timeline.

"Learn AI. Understand AI. Shape the Future."
— Susie Hala, IntelliGloss Press

Chapter 5

Designing AI-Smart Assignments

The emergence of generative artificial intelligence has forced educators to reconsider one of the most fundamental components of teaching: assessment. When a student can generate a structured essay or solve complex problems with a single prompt, traditional assignment formats require thoughtful redesign.

The goal of AI-smart assignment design is not to eliminate artificial intelligence from the learning environment. It is to ensure that assessment measures authentic understanding, reasoning, and growth.

This chapter provides a framework for designing assignments that maintain rigor, promote integrity, and strengthen higher-order thinking in an AI-influenced classroom.

Moving Beyond the "AI-Proof" Mindset

In the early stages of AI adoption, many educators attempt to create "AI-proof" assignments. While understandable, this approach often leads to restriction rather than innovation.

Artificial intelligence cannot be completely removed from students' environments. Attempting to design assignments solely around prevention can unintentionally narrow creativity and discourage exploration.

Instead of asking:

"How do I prevent AI use?"

A more productive question is:

"How do I design assignments that reveal student thinking?"

When thinking becomes visible, misuse becomes easier to detect and genuine learning becomes more measurable.

Principle 1: Make the Process Visible

Traditional assessments often evaluate only the final product. AI-smart design shifts attention to the process of learning.

Strategies include:

- Requiring brainstorming drafts

- Collecting annotated outlines

- Asking students to document revisions
- Incorporating reflection statements
- Conducting short in-class writing sessions

For example, a research paper may include:

1. Initial research questions
2. Source analysis notes
3. Outline submission
4. Draft with teacher feedback
5. Final revision
6. Reflection on learning process

This layered approach makes thinking observable.

Principle 2: Include Metacognitive Reflection

Reflection strengthens ownership of learning and reduces passive reliance on AI systems.

After completing an assignment, students can respond to prompts such as:

- What challenges did you encounter in developing your argument?
- How did you verify your sources?
- Did you use AI tools? If so, how?
- What revisions did you make after receiving feedback?

When students articulate their reasoning, teachers gain insight into comprehension depth.

Reflection also normalizes transparent AI use rather than secretive dependence.

Principle 3: Incorporate Oral Components

Oral explanation remains one of the most effective tools for assessing authentic understanding.

Options include:

- Brief presentation of key arguments
- Small group discussions
- Question-and-answer sessions
- One-on-one defense of thesis statements

When students explain their reasoning verbally, conceptual gaps become clearer. Oral components need not be lengthy to be effective. Even five minutes of explanation can reveal depth of comprehension.

Principle 4: Design Application-Focused Tasks

Artificial intelligence is highly effective at generating generic content. It is less effective when assignments require contextualized, personalized, or experiential application.

AI-smart assignments may include:

- Connecting concepts to local community issues
- Reflecting on personal experiences
- Applying theory to recent classroom discussions
- Conducting interviews or observational research
- Designing original projects tied to unique classroom themes

The more specific and contextual the assignment, the more authentic the student response becomes.

Principle 5: Teach Responsible AI Collaboration

Rather than banning AI tools outright, educators can teach structured collaboration.

For example:

Students may be instructed to:

- Generate an AI draft
- Critique its weaknesses
- Revise it significantly
- Document all modifications

This approach shifts AI from replacement to learning catalyst.

The emphasis remains on improvement, analysis, and revision—not on outsourcing effort.

AI-Resistant vs. AI-Enhanced Assignments

It can be helpful to distinguish between two types of assignment strategies.

AI-Resistant Assignments

Designed to emphasize independent reasoning.

Examples:

- Timed in-class writing
- Personal narrative reflection
- Live problem solving
- Group debate

These assignments ensure independent cognitive engagement.

AI-Enhanced Assignments

Structured to incorporate AI as a tool for exploration.

Examples:

- Comparing AI-generated summaries to textbook summaries
- Analyzing bias in AI responses
- Improving AI prompts for clarity
- Using AI to simulate historical dialogue

These assignments treat AI as an object of study rather than a shortcut.

Balanced instruction includes both types.

Establishing Classroom AI Guidelines

Assignment design should align with clear classroom expectations.

Educators may choose to establish:

- When AI use is permitted
- When it is restricted
- How AI assistance must be disclosed
- How citations should be handled

Transparency reduces confusion and reinforces trust.

Students benefit from knowing that expectations are structured rather than arbitrary.

Reclaiming Rigor

Artificial intelligence does not eliminate rigor. It shifts its definition.

Rigor in the AI era emphasizes:

- Analytical reasoning
- Evaluation of sources
- Original synthesis
- Ethical awareness
- Reflective thinking

Assignments designed with these goals in mind remain intellectually demanding regardless of technological advancement.

Designing for the Future

Artificial intelligence will continue evolving. However, the core educational objectives remain constant: to cultivate thinkers who can reason, question, and create responsibly.

AI-smart assignments do not lower standards. They clarify them.

When educators design assessments that prioritize thinking over formatting, the classroom becomes resilient to technological change.

In the next chapter, we translate these design principles into a structured 16-week implementation model that schools can adapt immediately.

"Learn AI. Understand AI. Shape the Future."
— Susie Hala, IntelliGloss Press

Chapter 6

AI-Enhanced Lesson Planning for Teachers

Artificial intelligence has the potential to reduce planning time, generate instructional ideas, and support differentiated learning. However, without clear structure, AI use in lesson design can feel overwhelming or unreliable.

The purpose of AI-enhanced lesson planning is not to replace teacher expertise. It is to support it.

Teachers remain instructional decision-makers. AI becomes an assistant—one that can generate drafts, suggest alternatives, and accelerate brainstorming. The professional judgment of the educator determines what is appropriate, accurate, and aligned with classroom goals.

This chapter outlines how teachers can use AI thoughtfully while preserving instructional quality.

Using AI as a Planning Assistant

Lesson planning requires significant time and cognitive effort. Artificial intelligence tools can assist by:

- Generating lesson outlines
- Creating sample discussion questions
- Producing vocabulary lists
- Suggesting project ideas
- Drafting formative assessment questions
- Providing alternative explanations of complex concepts

For example, a teacher preparing a lesson on climate change might request:

"Generate three discussion questions that promote critical thinking about the environmental impact of industrialization for a 10th-grade class."

The AI response may offer a starting point. The teacher then refines, edits, and adapts the questions to align with specific classroom needs.

The teacher remains the author. AI provides the draft.

Differentiation Support

One of the most valuable applications of AI in planning is differentiation.

Teachers can use AI to:

- Simplify reading passages for varied reading levels
- Generate enrichment extensions for advanced learners
- Create scaffolded question sets
- Produce multiple examples of a concept
- Translate vocabulary into student-friendly explanations

For instance, after designing a core lesson, a teacher may request:

"Rewrite this paragraph at a middle school reading level while maintaining key vocabulary terms."

The output provides a starting point that can be reviewed and refined.

Differentiation becomes more manageable when initial drafts are generated efficiently.

Generating Formative Assessments

AI tools can assist in creating:

- Exit tickets
- Short quizzes
- Practice questions
- Reflection prompts
- Case study scenarios

However, teachers should always review for:

- Accuracy
- Alignment with learning objectives
- Clarity of wording
- Bias or unintended assumptions

AI can accelerate drafting, but professional review ensures instructional integrity.

Refining Lesson Objectives

Clear learning objectives guide effective instruction.

Teachers may use AI to:

- Rephrase objectives for clarity
- Align objectives with skill categories
- Suggest measurable verbs
- Develop "I can" statements for students

For example:

"Convert this broad objective into three measurable learning outcomes for a 9th-grade history lesson."

The AI-generated suggestions can then be adjusted to meet district standards.

Scenario-Based Learning Design

Artificial intelligence can also assist in creating hypothetical scenarios or case studies that stimulate critical thinking.

For example:

- Simulated ethical dilemmas involving AI use

- Historical dialogue reconstructions
- Mock policy debates
- Real-world problem simulations

Teachers can use these generated scenarios as discussion starters or project foundations.

Again, the teacher reviews and edits to ensure contextual accuracy and developmental appropriateness.

Avoiding Overreliance

While AI offers efficiency, overreliance can reduce instructional intentionality.

Teachers should consider:

- Does this lesson reflect my instructional voice?
- Does it align with my students' needs?
- Have I verified all factual information?

- Does this activity support deeper thinking?

AI-generated material should be treated as a draft, not a final product.

Professional discernment remains central.

Ethical and Professional Considerations

When using AI for planning, educators should remain mindful of:

- Student data privacy
- District technology policies
- Accuracy of generated content
- Cultural sensitivity in language examples

AI tools should never replace careful professional judgment.

Transparency within professional learning communities can also support responsible integration. Teachers may benefit from sharing effective AI use strategies with colleagues.

Building Confidence Through Gradual Integration

Teachers new to AI-enhanced planning may begin with small steps:

- Generating discussion questions
- Drafting exit tickets

- Revising instructional explanations
- Brainstorming project ideas

Gradual experimentation builds familiarity and reduces uncertainty.

AI integration does not require immediate transformation. It can evolve alongside comfort and competence.

Preserving the Human Core of Teaching

Artificial intelligence can assist with efficiency, but it cannot replicate:

- Classroom rapport
- Emotional support
- Adaptive instruction in real time
- Professional intuition
- Cultural responsiveness

These human-centered elements remain irreplaceable.

When used thoughtfully, AI enhances preparation time while preserving relational teaching.

The AI-ready classroom does not automate the educator. It strengthens the educator's capacity to focus on meaningful interaction, creativity, and student engagement.

In the next chapter, we present a structured 16-week implementation model that translates these strategies into a cohesive AI literacy unit for grades 6–12.

"Learn AI. Understand AI. Shape the Future."
— Susie Hala, IntelliGloss Press

Chapter 7

Sample Classroom AI Policy and Parent Communication Framework

Artificial intelligence integration in education requires clarity, transparency, and trust. While AI tools can enhance learning, unstructured use can create confusion for students and families. A clearly defined classroom AI policy helps establish expectations, maintain academic integrity, and promote responsible engagement.

This chapter provides a model framework that schools and teachers may adapt to align with district policies and local guidelines.

Why a Classroom AI Policy Matters

When expectations are unclear, students may:

- Use AI tools without understanding boundaries
- Assume all AI assistance is acceptable
- Fear disclosure of AI use
- Misinterpret academic integrity guidelines

A classroom policy accomplishes three goals:

1. It clarifies when AI use is appropriate.
2. It reinforces accountability and transparency.
3. It protects both students and educators.

Artificial intelligence should not exist in a gray area. Clear guidelines reduce uncertainty and build confidence.

Core Principles of a Classroom AI Policy An

effective AI classroom policy should include:

- Transparency
- Responsibility
- Academic integrity
- Data awareness
- Teacher discretion

The goal is not restriction for its own sake, but structured engagement.

Sample Classroom AI Usage Guidelines

The following model may be adapted to individual classrooms or schools:

1. Permitted Uses

Students may use approved AI tools for:

- Brainstorming ideas
- Generating outlines

- Clarifying vocabulary
- Reviewing grammar
- Exploring multiple perspectives
- Practicing skill development

Use must be disclosed when required.

2. Restricted Uses

AI tools may not be used to:

- Complete assignments designated as independent work
- Generate final essays without teacher approval
- Replace required research processes
- Fabricate citations or references
- Bypass critical thinking expectations

Teachers will clearly indicate when AI assistance is prohibited.

3. Disclosure Requirement

When AI tools are used, students must:

- Identify the tool used
- Briefly describe how it was used
- Explain how they revised or improved the output

Example disclosure statement:

"I used an AI tool to generate an initial outline. I revised the structure, added original analysis, and verified all sources independently."

Transparency reinforces academic integrity.

4. Verification Responsibility

Students are responsible for:

- Fact-checking AI-generated information
- Confirming source accuracy
- Correcting errors or inconsistencies
- Ensuring the final submission reflects their understanding

AI-generated content does not exempt students from accountability.

5. Privacy Awareness

Students should avoid:

- Entering personal identifying information
- Sharing confidential data
- Uploading sensitive school materials without permission

Teachers should consult district policies regarding approved platforms.

Sample AI Academic Integrity Statement

Teachers may include a short statement in their syllabus:

"This classroom recognizes that artificial intelligence tools are part of the modern learning environment. Students may use AI tools when explicitly permitted. However, all submitted work must reflect the student's own understanding and reasoning. Undisclosed or inappropriate AI use may be considered a violation of academic integrity policies."

This language reinforces clarity without fear-based messaging.

Parent Communication Template

Transparent communication with families builds trust and reduces misunderstandings.

Below is a sample letter that teachers may adapt:

Sample Parent Communication

Dear Families,

As part of our commitment to preparing students for the evolving digital world, our classroom will be engaging in structured discussions about artificial intelligence (AI). Students may encounter AI tools in academic and everyday settings. Our goal is to help them understand how these tools work and how to use them responsibly.

In our classroom:

- AI use will be clearly guided and supervised.
- Students will be taught to evaluate AI responses critically.
- Academic integrity expectations remain in place.
- Transparency and responsible use are required.

Artificial intelligence will not replace student thinking. Instead, we will use it as a case study to strengthen reasoning, ethics, and critical analysis skills. If you have any questions about how AI is being addressed in class, please feel free to contact me.

Sincerely,

Teacher Name

This communication emphasizes education rather than promotion.

Adapting Policy at the School Level

Schools may expand classroom policies into broader guidelines that include:

- Approved platforms
- Teacher training expectations
- Professional development plans
- Data compliance measures
- Equity considerations

AI literacy initiatives should align with district-level decisions and legal frameworks.

Building a Culture of Responsible AI Use

Policy alone is not enough. Classroom culture must reinforce:

- Honesty
- Curiosity
- Accountability
- Reflection

When students understand expectations and feel safe discussing AI use openly, the classroom environment becomes collaborative rather than adversarial.

Responsible AI integration begins with clarity, not control.

In the next chapter, we present a 16-week sample AI literacy unit that schools can adapt for immediate implementation.

"Learn AI. Understand AI. Shape the Future."
— Susie Hala, IntelliGloss Press

Chapter 8

A 16-Week AI Literacy Implementation Model for Grades 6–12

Artificial intelligence literacy becomes meaningful when it moves beyond discussion and into structured instruction. While schools may vary in scheduling models, course structures, and instructional time, a twelve-week framework provides a manageable entry point for implementation. This timeline is long enough to build conceptual depth and short enough to integrate within an existing semester.

The purpose of this model is not to mandate uniformity. Rather, it offers a flexible structure that schools may adapt according to grade level, subject area, and district priorities. The framework aligns with the four pillars introduced earlier: conceptual understanding, responsible use, critical evaluation, and creative application.

Weeks 1–2: Foundations of Artificial Intelligence

The first phase establishes conceptual clarity. Students are introduced to the definition of artificial intelligence, the distinction between narrow AI and theoretical general AI, and the role of data and algorithms in generating responses. Instruction during this phase focuses on demystifying AI.

Teachers guide students through discussions about how AI systems generate language, recognize patterns, and produce outputs based on probability rather than human comprehension. Classroom activities may include analyzing examples of AI-generated text and identifying characteristics such as fluency, repetition, or subtle inaccuracies.

By the end of this phase, students should understand that AI simulates intelligence through pattern recognition and statistical prediction. This foundational knowledge reduces both overestimation and fear.

Weeks 3–4: Responsible Use and Academic Integrity

The second phase addresses ethical and responsible engagement. Students examine appropriate and inappropriate uses of AI tools within academic contexts. Teachers introduce classroom policy guidelines and facilitate discussion about transparency, disclosure, and verification.

Students may analyze hypothetical scenarios involving AI misuse and debate appropriate responses. Emphasis is placed on the idea that AI is a tool requiring accountability. Lessons also address privacy awareness and data considerations, helping students recognize the importance of protecting personal information.

By the end of this phase, students should demonstrate an understanding of when AI assistance may be permitted and how to disclose its use appropriately.

Weeks 5–6: Critical Evaluation of AI Outputs

During this phase, instruction shifts toward analytical thinking. Students examine AI-generated responses for accuracy, bias, and logical coherence. Teachers model how to verify information using credible sources and encourage skepticism toward overly confident outputs.

Classroom exercises may include comparing AI-generated essays to peer-written essays, identifying strengths and weaknesses in each. Students practice revising AI drafts to improve clarity, depth, and originality. The focus is not on accepting AI output but on evaluating and refining it.

This stage strengthens research skills and reinforces the importance of independent reasoning.

Weeks 7–8: AI-Smart Assignment Design and Application

In this phase, students begin engaging with AI more actively under structured conditions. Teachers design assignments that either limit AI use for independent assessment or incorporate AI collaboration intentionally.

Students may be tasked with generating an AI-assisted outline and then expanding it through independent analysis. Reflection components require students to describe how they revised AI suggestions and what intellectual contributions they added.

Instruction emphasizes that AI can assist brainstorming and exploration but cannot replace human judgment or synthesis.

Weeks 9–10: Project-Based Exploration

Students apply their knowledge in more complex, interdisciplinary tasks. For example, they may research how AI is used in healthcare, environmental science, journalism, or creative arts. Projects may involve presentations, written reports, or collaborative debates.

At this stage, students are encouraged to examine broader societal implications, including ethical considerations, equity concerns, and workforce transformation. Teachers facilitate structured dialogue that connects AI literacy to real-world applications.

The objective is to deepen understanding while reinforcing evaluation and responsible engagement.

Weeks 11–16: Reflection and Synthesis

The final phase consolidates learning. Students reflect on how their understanding of artificial intelligence has evolved. Teachers may assign capstone reflections, presentations, or comparative analyses that require students to synthesize concepts from earlier weeks.

Students articulate what AI can and cannot do, describe responsible usage practices, and evaluate its role in academic and professional contexts. Reflection may include written analysis, oral defense, or structured classroom discussion.

By the conclusion of the twelve-week model, students should demonstrate:

- Conceptual understanding of AI systems
- Awareness of ethical considerations
- Ability to critically evaluate outputs
- Capacity to use AI responsibly and reflectively

Flexibility Across Grade Levels

This model may be adapted based on developmental readiness. In middle school settings, instruction may emphasize foundational concepts and guided activities. In high school classrooms, deeper exploration of machine learning principles, ethical debates, and independent research projects may be appropriate.

The framework is modular rather than rigid. Schools may integrate it within technology courses, English language arts, social studies, or interdisciplinary programs.

Sustaining the Model

Artificial intelligence will continue evolving. However, the competencies cultivated in this 16 week framework remain durable. Understanding, responsibility, evaluation, and application are transferable skills that extend beyond specific platforms.

When implemented thoughtfully, this model supports not only AI literacy but also critical thinking, ethical reasoning, and academic integrity.

The next chapter will introduce assessment rubrics and evaluation tools that align with this implementation framework and help educators measure student growth consistently.

"Learn AI. Understand AI. Shape the Future."
— Susie Hala, IntelliGloss Press

Chapter 9

Assessment Rubrics and Evaluation Tools for AI Literacy

Artificial intelligence literacy cannot rely solely on discussion or exploration. Like all meaningful instruction, it requires structured evaluation. Assessment in the AI-ready classroom should measure not only content knowledge but also reasoning, ethical awareness, and reflective capacity.

Traditional grading models often emphasize final products. In an AI-influenced environment, however, evaluation must account for process, discernment, and transparency. This chapter presents assessment structures aligned with the IntelliGloss AI Literacy Framework, ensuring that student growth is measurable and consistent.

Assessing Conceptual Understanding

The first pillar of AI literacy involves foundational knowledge. Students must demonstrate accurate comprehension of how artificial intelligence systems function.

Assessment of conceptual understanding may include written explanations, short analytical responses, or structured classroom discussions. Students should be able to articulate:

- The difference between narrow AI and theoretical general AI
- How AI generates outputs through pattern recognition
- Why AI systems may produce inaccuracies
- The limitations of machine learning models

Evaluation criteria should emphasize clarity of explanation, accuracy of terminology, and ability to distinguish between common misconceptions and factual understanding.

Rather than memorization, assessment should focus on conceptual reasoning.

Assessing Responsible Use

Responsible engagement with AI requires observable accountability. Teachers may evaluate this competency through disclosure statements, reflection components, and adherence to classroom guidelines.

Students should demonstrate:

- Transparent reporting of AI use
- Awareness of academic integrity expectations

- Ability to identify when AI assistance is appropriate
- Understanding of privacy considerations

Assessment may include written reflections explaining how AI was used in a project and how the student verified or revised generated content.

The goal is not punitive enforcement but reinforcement of ethical habits.

Assessing Critical Evaluation Skills

Critical evaluation is one of the most important competencies in AI literacy. Students must show the ability to analyze AI-generated responses thoughtfully.

Evaluation tools may measure:

- Identification of factual inaccuracies
- Recognition of bias or incomplete perspectives
- Ability to compare AI responses with credible sources
- Depth of analytical revision

For example, a teacher might provide an AI-generated paragraph containing subtle errors. Students would then identify weaknesses and propose corrections. Grading criteria would focus on analytical reasoning rather than surface-level edits.

This form of assessment strengthens discernment and intellectual independence.

Assessing Creative and Collaborative Application

When students use AI as a structured collaborator, evaluation must distinguish between automation and synthesis.

Teachers may assess:

- Original contributions beyond AI-generated drafts
- Depth of revision and refinement
- Integration of personal insight
- Evidence of independent thinking

Students should be able to describe how they modified AI suggestions and explain the reasoning behind those revisions.

Evaluation of creative application ensures that AI enhances, rather than replaces, student cognition.

Developing a Balanced Rubric Model

An effective AI literacy rubric may include four performance categories aligned with the framework:

Conceptual Accuracy
Responsible Engagement
Critical Evaluation
Reflective Insight

Each category can be evaluated across developmental levels such as emerging, developing, proficient, and advanced.

For example, a student demonstrating proficiency in critical evaluation would consistently identify inaccuracies in AI output and provide evidence-based corrections. An advanced student might also recognize subtle bias patterns and propose alternative perspectives.

The rubric should prioritize reasoning quality over stylistic polish.

Incorporating Formative Assessment

Summative evaluation is important, but formative assessment plays a critical role in skill development. Teachers may use:

- Exit reflections
- Short analytical prompts
- Peer review exercises
- Classroom discussions
- Revision conferences

These tools allow educators to monitor progress before final evaluation occurs.

Formative assessment reinforces growth and reduces anxiety surrounding AI integration.

Maintaining Fairness and Consistency

Because AI literacy is an emerging domain, transparency in grading is essential. Students should understand how their work will be evaluated and how AI use factors into assessment.

Teachers may choose to include grading statements such as:

"Evaluation prioritizes evidence of independent reasoning, critical analysis, and responsible AI engagement."

Clear expectations promote fairness and reduce misunderstandings.

Measuring Growth Over Time

AI literacy is developmental. Students may initially struggle to critique AI responses or articulate limitations. Growth becomes visible when students begin questioning outputs independently and demonstrating reflective thinking.

Teachers may consider using pre- and post-unit reflections to measure conceptual progress. Comparing early definitions of AI with later explanations can reveal increased clarity and nuance.

Assessment should capture this evolution.

Aligning Evaluation with Educational Goals

Ultimately, assessment in the AI-ready classroom should reinforce broader educational objectives:

- Intellectual independence
- Ethical awareness
- Analytical rigor
- Responsible digital citizenship

Artificial intelligence literacy is not an isolated skill. It intersects with research, writing, problem-solving, and civic engagement.

When evaluation tools reflect these priorities, AI instruction strengthens overall academic development.

The next chapter will address standards alignment and implementation guidance for schools seeking to integrate AI literacy within existing curriculum structures.

"Learn AI. Understand AI. Shape the Future."
— Susie Hala, IntelliGloss Press

Chapter 10

Assessment and Evaluation in the AI-Ready Classroom

Assessment in the AI-ready classroom must evolve alongside instructional practice. When artificial intelligence becomes part of the learning environment, traditional evaluation methods that focus exclusively on final products are no longer sufficient. Educators must design assessment models that reveal reasoning, ethical awareness, and independent thought. The purpose of evaluation is not to detect technology use but to measure intellectual development.

Effective assessment begins with conceptual understanding. Students should demonstrate accurate comprehension of how artificial intelligence systems function. Rather than recalling isolated definitions, learners must be able to explain in their own words how AI generates responses through pattern recognition and statistical prediction. They should understand that such systems do not possess human awareness or intention and that apparent fluency does not guarantee accuracy. Assessment of this understanding may occur through written explanations, guided discussion, or analytical responses that require students to clarify misconceptions. The emphasis should remain on clarity of reasoning rather than memorization of terminology.

Responsible engagement constitutes another dimension of evaluation. As artificial intelligence tools become accessible, accountability becomes a measurable skill. Students must demonstrate transparency regarding their use of AI assistance and reflect on how such tools contributed to their work. Evaluation in this area should consider whether students disclose AI usage appropriately, whether they verify generated information independently, and whether their final submission reflects authentic understanding rather than unexamined output. Ethical awareness becomes visible when students articulate why certain uses are appropriate in some contexts but not in others.

Critical evaluation is perhaps the most distinctive feature of assessment in an AI-influenced classroom. Students should be able to analyze AI-generated responses for accuracy, bias, and coherence. When presented with an automated paragraph, learners must identify unsupported claims, vague reasoning, or incomplete analysis. More advanced students may recognize subtle patterns of bias or structural weaknesses that require revision. Evaluation should reward depth of analysis and evidence-based correction rather than surface-level editing. This dimension of assessment strengthens intellectual independence and reinforces the habit of questioning authoritative-sounding outputs.

Creative and collaborative application also requires careful measurement. When students use AI tools as part of a structured assignment, the focus of evaluation shifts to synthesis and refinement. Teachers should examine how students transform initial drafts into substantive work that reflects personal insight and analytical depth. Growth becomes evident when students demonstrate the ability to move beyond generated suggestions and produce nuanced, original contributions. Reflection statements can support this evaluation by revealing the reasoning behind revisions and choices.

Balanced assessment in the AI-ready classroom therefore integrates conceptual clarity, ethical responsibility, analytical critique, and reflective synthesis. These elements align with broader educational goals and ensure that artificial intelligence enhances rather than diminishes academic rigor.

As with any emerging instructional domain, fairness and transparency remain essential. Students should understand how their work will be evaluated and how AI usage factors into grading decisions. Clear communication reduces anxiety and supports a culture of integrity. Over time, assessment becomes not merely a measurement tool but a reinforcement mechanism for responsible and thoughtful engagement with intelligent systems.

Chapter 11

Standards Alignment and Curriculum Integration

Artificial intelligence literacy does not exist outside the framework of established educational goals. Although AI is an emerging topic, the competencies it develops—critical thinking, ethical reasoning, analytical evaluation, and responsible digital citizenship—are already embedded within existing academic standards. The integration of AI literacy, therefore, does not require a departure from traditional educational objectives. Rather, it strengthens them.

Across grade levels, educational standards consistently emphasize reasoning, problem-solving, communication, and evidence-based analysis. AI literacy naturally supports these priorities. When students analyze AI-generated responses, they practice evaluating sources and identifying logical inconsistencies. When they reflect on responsible use, they engage in ethical reasoning and civic awareness. When they revise AI-assisted drafts, they demonstrate synthesis and ownership of ideas.

In language arts classrooms, AI literacy connects directly to reading comprehension, argumentative writing, and research standards. Students comparing AI-generated summaries to primary texts deepen their ability to identify nuance and bias. Structured reflection reinforces metacognitive awareness, a skill widely recognized as essential for academic growth.

In social studies contexts, discussions about artificial intelligence intersect with civic responsibility, technological impact, and public policy. Students examining how AI influences media, employment, and decision-making processes develop analytical frameworks applicable to democratic participation.

Science and technology courses can integrate AI literacy through exploration of machine learning concepts, data interpretation, and ethical implications of automation. Students studying algorithmic bias or data limitations engage directly with scientific reasoning and evidence evaluation.

Mathematics classrooms may incorporate AI discussions by examining statistical prediction models, probability, and pattern recognition. These connections reinforce quantitative reasoning while contextualizing abstract concepts within real-world applications.

Beyond subject-specific alignment, AI literacy supports broader digital citizenship initiatives. Educational institutions have long emphasized responsible technology use. Artificial intelligence represents an extension of this commitment. Students who understand how AI systems function are better equipped to evaluate online information, recognize misinformation, and protect personal data.

Institutional implementation requires thoughtful planning. Schools may choose to integrate AI literacy within an existing course, embed it across disciplines, or offer a dedicated module.

Regardless of structure, consistency is essential. Faculty collaboration strengthens alignment and ensures shared expectations regarding AI use and academic integrity.

Professional development plays a critical role in successful implementation. Educators must feel supported rather than pressured. Gradual integration, beginning with foundational understanding and expanding toward applied practice, reduces uncertainty and builds confidence. Administrators can foster this growth by encouraging collaborative dialogue and sharing effective instructional strategies.

Communication with families further reinforces institutional trust. When schools articulate that AI literacy focuses on ethical reasoning, analytical rigor, and responsible engagement, stakeholders recognize its educational value. Transparency transforms uncertainty into partnership.

The long-term sustainability of AI literacy depends on principle-based integration rather than tool-based dependence. Technologies will evolve, but core competencies endure. By anchoring AI instruction in reasoning, responsibility, and reflective thinking, schools create a model adaptable to future innovation.

Artificial intelligence does not require the reinvention of educational standards. It invites their application in new contexts. When thoughtfully implemented, AI literacy strengthens existing academic goals and prepares students for informed participation in a technologically complex world.

"Learn AI. Understand AI. Shape the Future."
— Susie Hala, IntelliGloss Press

Chapter 12

Institutional Leadership and Administrative Guidance

The successful integration of artificial intelligence literacy within a school requires more than classroom enthusiasm. Sustainable implementation depends on coordinated planning, professional collaboration, and administrative clarity. While individual educators may begin experimenting with AI-informed instruction independently, long-term effectiveness emerges when schools adopt a structured, institution-wide approach.

Implementation does not require immediate transformation. In fact, gradual and deliberate adoption often produces stronger outcomes. Schools that move thoughtfully can evaluate impact, refine expectations, and build confidence among faculty and families.

Establishing a Shared Understanding

Before instructional changes occur, administrators and faculty benefit from establishing a shared conceptual foundation. Artificial intelligence can evoke strong reactions ranging from excitement to skepticism. Clear, measured discussion reduces misinformation and prevents unnecessary polarization.

Initial faculty conversations may focus on clarifying what artificial intelligence is and is not. When educators understand that AI systems generate responses through pattern recognition rather than independent reasoning, anxiety often diminishes. Shared understanding allows instructional decisions to be guided by clarity rather than fear.

Leadership teams may consider dedicating professional learning sessions to exploring foundational AI concepts, reviewing classroom policy guidelines, and discussing instructional opportunities aligned with existing curriculum goals.

Defining Scope and Placement

Schools must determine how AI literacy will fit within their instructional structure. Several implementation models are possible.

Some institutions may choose to integrate AI literacy within existing subject areas such as English language arts, social studies, technology courses, or interdisciplinary programs. In this approach, AI serves as a contextual lens that strengthens established learning objectives.

Other schools may adopt a short-term module model, embedding a structured twelve-week unit within a semester. This model allows concentrated exploration without requiring a full course redesign.

A third approach involves offering an elective course focused on digital literacy and artificial intelligence. This model provides depth while allowing voluntary participation.

No single model is universally superior. Effective implementation depends on scheduling flexibility, faculty readiness, and institutional priorities.

Supporting Faculty Development

Professional development plays a critical role in building institutional confidence. Educators must feel supported as they navigate evolving technologies. Implementation should emphasize growth rather than evaluation.

Effective professional development may include collaborative workshops in which teachers explore AI tools together, analyze potential classroom applications, and discuss ethical considerations. Peer dialogue often reduces uncertainty more effectively than isolated training sessions.

Administrators should encourage gradual experimentation. Faculty members may begin by integrating AI-enhanced lesson planning strategies or by redesigning a single assignment to incorporate reflective AI evaluation. Small successes build collective momentum.

Aligning Policy and Practice

Clear institutional policy provides stability during implementation. Schools may develop guidelines that outline appropriate student use, disclosure expectations, and data protection considerations. These policies should remain adaptable as technologies evolve.

Consistency across classrooms strengthens clarity for students. When expectations vary significantly between courses, confusion increases. Faculty collaboration ensures that AI usage standards are coherent while still allowing instructional autonomy.

Regular policy review allows schools to adjust guidelines in response to new developments or feedback.

Engaging Families and Community Stakeholders

Artificial intelligence is frequently discussed in public media, often in exaggerated terms. Transparent communication with families reduces misunderstanding and fosters trust.

Schools may choose to host informational sessions explaining how AI literacy supports critical thinking and responsible digital citizenship. Providing clear examples of classroom applications reassures stakeholders that instruction is structured and educationally grounded.

When families understand that AI literacy emphasizes reasoning and ethical awareness, resistance often decreases.

Monitoring and Reflection

Implementation should include ongoing evaluation. Administrators and teachers may gather feedback from students, review assessment outcomes, and discuss instructional adjustments.

Reflection questions might include:

- Are students demonstrating improved critical evaluation skills?
- Do students understand responsible AI use expectations?
- Are faculty members comfortable integrating AI discussions into lessons?
- Does the current model align with school goals?

Periodic review ensures that AI literacy remains aligned with institutional mission.

Building a Culture of Thoughtful Adaptation

Artificial intelligence will continue evolving. Schools that adopt a principle-based approach—focused on understanding, responsibility, and critical thinking—are better positioned to adapt to future changes.

Implementation should not aim for technological perfection. It should aim for intellectual preparedness.

When institutions approach AI literacy as an extension of existing educational values rather than a disruptive replacement, integration becomes manageable and sustainable.

The final chapter of this guide reflects on the broader future of teaching and learning in an era shaped by intelligent systems and reaffirms the enduring centrality of human educators.

Chapter 13

Teaching Forward: Human Leadership in an AI-Influenced Era

Artificial intelligence represents a significant technological shift, but it does not redefine the fundamental purpose of education. Schools exist to cultivate reasoning, character, curiosity, and civic responsibility. These goals remain unchanged, even as tools evolve.

The presence of AI in classrooms invites adaptation, not abandonment. It challenges educators to clarify learning objectives, strengthen assessment design, and deepen ethical conversation. In doing so, it reinforces the central role of the teacher.

Throughout history, education has responded to technological change. The introduction of calculators altered mathematics instruction but did not eliminate the need for mathematical reasoning. The rise of the internet transformed research practices but increased the importance of source evaluation. Digital communication reshaped literacy while heightening the need for critical analysis.

Artificial intelligence follows this pattern. It does not remove the necessity of thinking; it demands more sophisticated thinking.

When students encounter AI-generated responses, they must learn to question accuracy, identify bias, and evaluate evidence. When they collaborate with intelligent systems, they must refine, revise, and synthesize. When they navigate ethical dilemmas related to automation, privacy, and decision-making, they engage in civic reasoning.

These are not new educational goals. They are enduring ones expressed in a new context.

Teachers remain the architects of learning environments. Artificial intelligence may assist with drafting, brainstorming, or simulation, but it cannot cultivate empathy, discern when a student is discouraged, or adapt instruction in response to subtle classroom dynamics. It cannot model integrity or demonstrate intellectual humility.

The AI-ready classroom is not defined by the presence of technology. It is defined by intentional guidance. Educators who understand how AI systems function can frame them as objects of study rather than unquestioned authorities. Students who learn to engage critically with intelligent systems develop habits of discernment that extend beyond the classroom.

Institutional leadership also plays a vital role. Schools that approach AI literacy through structured frameworks, thoughtful policy, and collaborative dialogue create stability during periods of change. Gradual implementation allows reflection and refinement, preventing reactive decision-making.

Looking ahead, artificial intelligence will continue evolving. New tools will emerge, and capabilities will expand. However, the core competencies emphasized throughout this guide—conceptual understanding, responsible engagement, critical evaluation, and reflective application—remain durable.

Education has always prepared students for a world in motion. AI literacy is simply the latest extension of that responsibility.

The future of teaching is not automated. It is guided.

Human educators remain central—not because technology is limited, but because education is relational, ethical, and deeply human.

Artificial intelligence may influence how students access information, but it does not replace the need for wisdom in interpreting it. That responsibility remains in the hands of educators.

Teaching forward in an AI-influenced era requires clarity, confidence, and commitment to enduring principles. With thoughtful implementation, schools can transform uncertainty into opportunity and ensure that students graduate not only technologically capable, but intellectually prepared.

Chapter 14

The AI Knowledge Ladder: From Bits to Artificial Intelligence

Introduction

For many educators, artificial intelligence can feel like a complex and mysterious subject. News headlines often describe AI as a powerful technology capable of performing impressive tasks such as recognizing images, generating text, or analyzing large amounts of information. While these capabilities may seem advanced, every artificial intelligence system is built upon a set of simple digital foundations.

Understanding these foundations helps teachers explain AI in a clear and approachable way for students.

One useful way to understand artificial intelligence is through what we may call the **AI Knowledge Ladder**. This ladder shows how intelligent systems are built step by step, beginning with the smallest unit of digital information and progressing toward more advanced forms of understanding.

The ladder can be summarized in six stages:

Bits → Data → Information → Knowledge → Intelligence → Artificial Intelligence

By understanding how these layers connect, educators gain a framework that makes artificial intelligence easier to explain and teach.

The Foundation: Bits

At the very base of every digital system is a unit called a **bit**.

A bit, short for **binary digit**, is the smallest unit of information in computing. A bit can have only two possible values:

0 or 1

These two values represent electrical states inside a computer system. While a single bit carries only a tiny amount of information, computers combine billions or even trillions of bits together to store and process digital information.

Everything in modern computing—text, images, videos, sound, and software—is ultimately represented using combinations of bits.

For educators, this concept is important because it reminds us that even the most advanced technologies begin with very simple building blocks.

From Bits to Data

When many bits are combined together, they form **data**.

Data refers to raw digital content that computers can store and process. This may include numbers, words, images, sounds, or measurements collected from sensors.

For example:

A photograph stored on a smartphone

A message typed into a search engine

A video watched online

A temperature reading from a sensor

All of these are examples of data represented as digital bits inside a computer system.

Artificial intelligence systems rely heavily on data. The quality and quantity of data available to a system often determine how well that system performs.

From Data to Information

Data becomes more meaningful when it is organized and interpreted. When raw data is processed in a way that reveals patterns or meaning, it becomes **information**.

For instance, a collection of numbers may represent raw data. When those numbers are organized to show trends, comparisons, or relationships, they begin to provide information.

In the context of artificial intelligence, information helps machines detect patterns within large datasets. These patterns allow AI systems to identify relationships between inputs and outcomes.

For example, an AI system trained to recognize images of animals may analyze thousands of pictures. Through this process, the system identifies visual patterns that distinguish one animal from another.

From Information to Knowledge

As systems analyze information repeatedly, they begin to develop **knowledge**.

Knowledge refers to an understanding of patterns that can be used to make predictions or decisions.

For example, after analyzing many examples of handwriting, an AI system may learn to recognize letters and words. The system gains knowledge about what certain shapes and patterns represent.

In human learning, knowledge is built through study, observation, and experience. In artificial intelligence, knowledge is built through training data and algorithms that analyze patterns within that data.

This stage is where machine learning systems begin to demonstrate useful capabilities.

From Knowledge to Intelligence

Intelligence involves applying knowledge to solve problems, make decisions, or respond to new situations.

Humans demonstrate intelligence when they reason through problems, adapt to unfamiliar circumstances, or create new ideas.

Artificial intelligence attempts to replicate some aspects of this process by applying learned patterns to new inputs.

For example:

An AI system may analyze medical images to help detect signs of disease.

A navigation system may calculate the fastest route based on traffic conditions.

A recommendation system may suggest movies based on a viewer's preferences.

In each case, the system uses learned knowledge to produce useful results.

Artificial Intelligence

Artificial intelligence refers to computer systems designed to perform tasks that normally require aspects of human intelligence.

These tasks may include: recognizing

speech identifying images

translating languages predicting

outcomes generating text or

recommendations

Modern AI systems combine large datasets, powerful algorithms, and computing resources to analyze patterns and generate responses.

However, it is important for educators and students to understand that AI systems do not possess human awareness or independent understanding. They operate by analyzing patterns in data and applying learned rules.

Recognizing both the strengths and limitations of AI systems helps students develop a balanced perspective on intelligent technologies.

Why the AI Knowledge Ladder Matters for Teachers

The AI Knowledge Ladder provides educators with a simple framework for explaining artificial intelligence in the classroom.

By breaking AI into stages, teachers can help students understand that complex technologies are built from understandable components.

When students see that artificial intelligence begins with simple bits and gradually develops through layers of data and learning, the subject becomes less intimidating and more accessible.

This framework also encourages students to ask deeper questions about how digital systems work and how technology influences society.

Key Takeaways for Educators

- Artificial intelligence is built upon simple digital foundations.
- Bits are the smallest unit of information in computing.
- Large collections of bits form data used by computer systems.
- Organized data becomes information that reveals patterns.
- Repeated analysis of information leads to knowledge.

- Intelligence involves applying knowledge to solve problems.
- Artificial intelligence systems use these layers to perform complex tasks.

Understanding this progression allows educators to guide students through the world of artificial intelligence with greater confidence and clarity.

"Learn AI. Understand AI. Shape the Future."
— Susie Hala, IntelliGloss Press

Chapter 15

Understanding Bits, Bytes, Data, and AI Training

Introduction

Artificial intelligence systems may appear complex, but they are built on simple digital foundations. In order to understand how machines learn, it is helpful to begin with the smallest building blocks of computing.

Every digital technology—from a smartphone photo to an advanced AI model—ultimately begins with tiny units of information known as **bits**. These bits combine together to form **bytes**, which then store the **data** that computers analyze and process. When large collections of data are used to teach machines how to recognize patterns, the process is known as **AI training**.

For educators introducing artificial intelligence concepts to students, understanding these digital layers provides clarity. It allows teachers to explain that even the most advanced intelligent systems are built step by step from simple components.

Bits: The Smallest Unit of Digital Information

A **bit**, short for *binary digit*, is the smallest unit of information in computing.

A bit can have only two possible values:

0 or 1

These two values represent electrical states inside a computer system. One value may represent an "off" state and the other an "on" state. Although a single bit contains very little information, computers combine vast numbers of bits to represent all forms of digital content.

Text, images, sound, and video are all ultimately stored as patterns of bits inside computer memory and storage systems.

Understanding bits helps students realize that digital technology is built from simple foundations.

Bytes: Grouping Bits Together

Because a single bit carries only a small amount of information, computers group bits together into units called **bytes**.

A **byte** typically consists of **eight bits**.

This grouping allows computers to represent a larger range of values. For example, a single byte can represent numbers, letters, and symbols used in digital text.

As digital systems evolved, larger units were created to represent growing amounts of information. These include:

Kilobyte (KB) – approximately one thousand bytes

Megabyte (MB) – approximately one million bytes

Gigabyte (GB) – approximately one billion bytes

Terabyte (TB) – approximately one trillion bytes

Modern computer systems process enormous quantities of data measured in these larger units.

Data: The Digital Content Computers Use

When bits and bytes are organized to represent meaningful content, they form **data**.

Data can take many forms, including:

numbers stored in spreadsheets words

and sentences in digital documents

images captured by cameras audio

recordings and music files video

recordings sensor readings from devices

In artificial intelligence systems, data serves as the primary source of learning.

For example, a machine learning system designed to recognize images of animals must analyze thousands or even millions of pictures. Each image becomes part of the dataset that helps the system recognize patterns.

The quality and diversity of the data strongly influence how well an AI system performs.

Datasets: Large Collections of Data

When many pieces of data are grouped together for analysis, they form a **dataset**.

Datasets are essential to artificial intelligence because they provide the examples that machines study during the learning process.

For example:

A language model may be trained using large collections of books, articles, and online text. By analyzing these datasets, the system learns patterns in language, grammar, and context.

Similarly, an image-recognition system may train on millions of photographs labeled with descriptions. Through repeated exposure to these examples, the system learns to recognize objects, shapes, and features.

The larger and more carefully prepared the dataset, the more effectively the AI system can identify patterns.

AI Training: Teaching Machines to Learn Patterns

The process of teaching machines to recognize patterns in data is known as **AI training**.

During training, algorithms analyze datasets to identify relationships between inputs and outputs. The system adjusts its internal parameters repeatedly as it processes examples, gradually improving its ability to make predictions.

For instance, a machine learning model designed to recognize handwritten numbers may be trained using thousands of labeled examples. Over time, the system becomes better at recognizing the patterns that distinguish one number from another.

Although this process may seem similar to human learning, it operates through mathematical optimization rather than human understanding.

AI systems do not truly "know" concepts in the way humans do. Instead, they identify statistical relationships within data.

Why This Matters for Teachers

Understanding bits, bytes, data, and AI training helps educators explain artificial intelligence in a structured and accessible way.

Rather than presenting AI as a mysterious or magical technology, teachers can guide students through the logical steps that lead to intelligent systems.

Students can see that:

digital information begins with bits bits

combine to form bytes bytes store data large

datasets provide examples algorithms analyze

those datasets training allows machines to

recognize patterns

This layered understanding builds confidence and encourages deeper curiosity about how modern technology works.

Key Takeaways for Educators

- A **bit** is the smallest unit of digital information and can represent either 0 or 1.
- Eight bits typically form a **byte**, which stores larger pieces of information.
- **Data** represents digital content such as text, images, and numbers.
- Large collections of data are called **datasets**.
- **AI training** occurs when algorithms analyze datasets to learn patterns.
- Understanding these layers helps teachers explain artificial intelligence clearly.

Chapter 16: The Complete AI School Model — Combining AIPowered Learning with AI Understanding

Introduction

Education is undergoing a profound transformation. Artificial intelligence is no longer a future concept—it is actively reshaping how instruction is delivered, how students learn, and how schools operate.

Across the education landscape, new models are emerging that use artificial intelligence to personalize learning, accelerate academic progress, and improve engagement. These systems challenge the traditional structure of the classroom by shifting away from standardized instruction toward adaptive, student-centered experiences.

At the same time, a critical need is becoming increasingly clear: students must not only use artificial intelligence, but also understand how it works.

This chapter presents a unified framework that combines these two essential directions—AIpowered learning and AI literacy—into a complete and practical model for modern schools.

Two Transformations Shaping Education

Education is currently being shaped by two powerful and interconnected transformations.

The first is the rise of AI-powered learning systems. These systems use artificial intelligence to continuously assess student performance and adjust instruction in real time. As a result, students can move at their own pace, receiving personalized support and advancing more efficiently through academic material.

The second transformation is the growing importance of AI literacy. As students interact with intelligent systems daily, it becomes essential for them to understand how these systems function, how decisions are made, and what limitations exist.

These two transformations are often treated separately. However, when combined, they form the foundation of a complete and future-ready educational model.

AI-Powered Learning: Efficiency and Personalization

AI-powered learning systems are designed to optimize how instruction is delivered.

Through continuous data analysis, these systems identify student strengths, gaps, and learning patterns. Instruction is then adjusted dynamically, ensuring that each learner receives content that matches their level of understanding.

This approach reduces the inefficiencies of traditional classrooms, where all students move at the same pace regardless of individual needs. It also increases engagement by providing appropriately challenging material.

In many implementations, core academic learning can be completed in a more focused time frame, allowing schools to rethink how the rest of the day is used.

AI Literacy: Building Understanding and Awareness

While AI-powered systems improve learning efficiency, AI literacy focuses on developing understanding.

AI literacy equips students with knowledge about the foundational systems behind artificial intelligence. This includes concepts such as data, algorithms, neural networks, tokens, parameters, infrastructure, and energy.

More importantly, AI literacy helps students develop critical thinking. They learn to question outputs, recognize limitations, and understand the broader impact of AI on society.

Without this layer of understanding, students may become dependent on tools without the ability to evaluate them.

Closing the Gap Between Use and Understanding

One of the most important challenges in modern education is the gap between using technology and understanding it.

Students can quickly become proficient at interacting with AI systems. However, without structured learning, they may not understand how those systems generate results or what factors influence their behavior.

This gap can lead to overreliance, misunderstanding, and missed opportunities for deeper learning.

Closing this gap requires intentional design. Schools must create environments where students both use AI and study it.

The Unified Model: A Complete AI School Framework

The integration of AI-powered learning and AI literacy creates a complete educational framework.

In this model, artificial intelligence is used to enhance academic instruction while also serving as a subject of study. Students benefit from personalized learning experiences while developing a clear understanding of the systems they interact with.

This dual approach ensures that students are not only efficient learners, but also informed and responsible users of technology.

Implementation in Schools

For educators and school leaders, the question is not whether this model is possible, but how it can be implemented effectively.

A practical approach is to structure the school day in two complementary phases.

The first phase focuses on AI-supported academic learning. During this time, students engage with adaptive systems that personalize instruction and allow for efficient mastery of core subjects.

The second phase focuses on AI literacy. Using structured curriculum resources, students explore how artificial intelligence works, analyze real-world applications, and engage in discussions about ethics and responsibility.

This structure does not require a complete redesign of the school system. It can be introduced gradually, starting with pilot programs, elective courses, or integrated modules.

The Role of Teachers in the New Model

Teachers remain central to this model, but their role evolves.

Rather than focusing primarily on delivering content, educators guide understanding, facilitate discussions, and support critical thinking. They help students connect their experiences with AI tools to the underlying concepts that drive those tools.

Teachers also play a key role in fostering responsible use of technology, ensuring that students understand both the capabilities and limitations of AI.

This shift allows educators to focus on higher-value interactions while AI supports routine instructional tasks.

Benefits for Schools and Districts

The combined model offers several important advantages for schools and districts.

It aligns education with current technological realities, ensuring that students are prepared for a world increasingly shaped by artificial intelligence.

It improves learning efficiency while maintaining depth and rigor.

It provides educators with a clear framework for integrating AI without losing the human element of teaching.

Most importantly, it positions schools as forward-thinking institutions that are proactively preparing students for the future.

Preparing for the Future of Education

The future of education will not be defined by technology alone, but by how that technology is used and understood.

Students who can both use and understand artificial intelligence will be better equipped to navigate complex challenges, adapt to change, and contribute meaningfully to society.

Schools that embrace this balanced approach will be better positioned to lead in an era of rapid technological advancement.

A New Standard for Modern Schools

The integration of AI-powered learning and AI literacy represents a new standard for education.

It is not a replacement for traditional teaching, but an evolution of it.

By combining efficiency with understanding, and innovation with responsibility, schools can create learning environments that are both effective and meaningful.

This is not simply a new model—it is a new direction for education.

Final Thought for Educators

Artificial intelligence is already in the classroom.

The question is no longer whether schools will adopt AI, but how they will adopt it.

Will students simply use these systems, or will they understand them?

The answer to that question will define the future of education.

Conclusion

Building an AI-Literate Generation

Artificial intelligence is no longer a distant concept reserved for research laboratories or technology companies. It is becoming part of everyday life, influencing how information is created, shared, and understood. From digital assistants and recommendation systems to medical research and transportation technologies, intelligent systems are quietly shaping the modern world.

Because of this transformation, education must evolve alongside technology. Schools have always played a critical role in preparing students for the future. Today, that preparation includes helping students understand the foundations of artificial intelligence.

AI literacy does not mean every student must become a programmer or computer scientist. Instead, it means students gain the ability to understand how intelligent systems work, recognize their limitations, and think critically about how technology influences society.

Throughout this book, educators have explored the foundations of artificial intelligence, the importance of AI literacy in schools, and practical strategies for teaching these concepts responsibly. Teachers have examined how AI can influence classroom learning, how to design thoughtful policies for its use, and how to guide students through discussions about technology, ethics, and the future.

The goal of AI education is not simply to teach technology. It is to empower students with understanding.

Students who develop AI literacy will be better prepared to evaluate digital information, ask thoughtful questions about technology, and participate responsibly in a rapidly evolving world. They will be able to distinguish between hype and reality, recognize both the opportunities and the limitations of intelligent systems, and contribute thoughtfully to conversations about the role of technology in society.

Educators stand at the center of this transformation. Teachers help translate complex ideas into meaningful learning experiences, guiding students toward understanding rather than confusion.

By introducing artificial intelligence education into classrooms, schools take an important step toward preparing the next generation of thinkers, innovators, and responsible citizens.

Artificial intelligence is shaping the future.

Education must help students understand it.

And with informed educators leading the way, classrooms can become places where curiosity, responsibility, and knowledge come together to build an AI-literate generation.

Conclusion
Building an AI-Literate Generation

"Learn AI. Understand AI. Shape the Future."
— Susie Hala, IntelliGloss Press

GLOSSARY

Letter A

1. Access Time

The amount of time it takes for an AI system or computer to locate and retrieve data from memory or storage. Access time affects how quickly megabytes of data can be used during AI processing.

2. Activation Data

Intermediate data generated inside an AI model while it is running. Activation data is stored temporarily in memory and often measured in megabytes during model execution.

3. Adaptive Storage

A storage approach that automatically adjusts how data is stored or moved based on usage patterns. In AI systems, adaptive storage helps manage large volumes of megabytes efficiently.

Artificial General Intelligence (AGI)

Definition
Artificial General Intelligence (AGI) refers to a type of artificial intelligence that can understand, learn, and apply knowledge across a wide range of tasks at a level similar to human intelligence. Unlike narrow AI, which is designed for specific tasks, AGI can adapt to new situations, solve unfamiliar problems, and think more flexibly.

Key Characteristics

Ability to learn and apply knowledge across different subjects

General problem-solving skills similar to humans

Adaptability to new and unfamiliar tasks

Understanding of context, reasoning, and decision-making

Potential to transfer learning from one area to another

Simple Explanation (Student-Friendly)
AGI is an AI that can think and learn like a human, not just perform one specific task.

Example
A system that can write an essay, solve a math problem, understand emotions in a conversation, and learn a completely new skill without being specifically programmed for each one.

Applications (Future-Oriented)

Advanced personal assistants

Scientific research and discovery

Complex decision-making systems

Autonomous systems that can operate in many environments

Important Note
AGI does not fully exist yet. Most AI systems today are examples of **Artificial Narrow Intelligence (ANI)**, meaning they are designed for specific tasks only.

Summary
Artificial General Intelligence represents the next major stage in AI development, where machines move beyond task-specific functions and begin to demonstrate flexible, human-like intelligence across many areas.

4. Algorithmic Data Size

The total amount of data an algorithm needs to operate effectively. In AI, this size is often measured in megabytes or larger units, depending on model complexity.

5. Allocation (Memory Allocation)

The process of assigning a specific number of megabytes in memory or storage for an AI task, model, or dataset.

6. Analytics Dataset

A structured collection of data used for analysis or AI training. The size of an analytics dataset is commonly measured in megabytes, gigabytes, or more.

ANI (Artificial Narrow Intelligence)

Artificial Narrow Intelligence refers to AI systems designed to perform **one specific task or a narrow set of tasks** rather than general intelligence. ANI systems rely on data stored and processed in memory and storage measured in **megabytes, gigabytes, or more**. Examples include image recognition, speech-to-text, recommendation systems, and chatbots. ANI does not understand beyond its training scope, but it can process large amounts of data efficiently within its defined function.

7. Annotation Storage

The space required to store labels, tags, or explanations added to AI training data. These annotations contribute additional megabytes to total dataset size.

8. Artificial Neural Network (ANN) Size

The total memory footprint of a neural network, including weights, biases, and parameters. ANN size is often described in megabytes when discussing model efficiency.

ASCII (American Standard Code for Information Interchange)

Simple Definition (student-friendly)

ASCII is a system that computers use to represent letters, numbers, and symbols using binary (0s and 1s).

How It Works

Every character is assigned a number, and that number is stored in binary.

For example:

$A \rightarrow 65 \rightarrow 01000001$

$B \rightarrow 66 \rightarrow 01000010$

$a \rightarrow 97 \rightarrow 01100001$ 1

$\rightarrow 49 \rightarrow 00110001$

So when you type on a keyboard, the computer doesn't see letters — it sees **binary codes based on ASCII**

ASCII Uses 8 Bits (1 Byte)

Each character = **8 bits**

Example:

01000001

This binary represents the letter **A**

Why ASCII Matters

ASCII connects directly to your **Bit Family + Data Family**:

Bits → form binary

Binary → forms characters (ASCII)

Characters → form words

Words → form data

Data → feeds AI

This is the **foundation of all computing and AI**

Simple Way to Explain to Students You

can say:

"ASCII is like a translation system that turns letters into numbers so computers can understand them."

Artificial Superintelligence (ASI)

Definition
Artificial Superintelligence (ASI) refers to a level of artificial intelligence that surpasses human intelligence in all areas, including reasoning, problem-solving, creativity, decision-making, and emotional understanding. ASI would be capable of performing tasks far beyond the abilities of the most advanced human minds.

Key Characteristics

Intelligence that exceeds human capability in every domain

Ability to solve extremely complex global problems

Rapid learning and self-improvement

Advanced reasoning, creativity, and innovation

Potential to operate independently at a high level

Simple Explanation (Student-Friendly)
ASI is an AI that is smarter than humans in every way.

Example (Conceptual)
A system that can discover new scientific breakthroughs, solve global challenges like climate change, design advanced technologies, and continuously improve itself without human guidance.

Applications (Future-Oriented)

Global problem-solving (healthcare, climate, energy)

Advanced scientific discovery

Autonomous decision-making at large scales

Creation of new technologies beyond current human understanding

Important Note
ASI does not exist today. It is a theoretical future stage of artificial intelligence that researchers and experts continue to study and debate.

Challenges and Considerations

Ethical concerns about control and decision-making

Potential risks if not aligned with human values

Need for strong safety frameworks and governance

Questions about trust, responsibility, and impact on society

Summary
Artificial Superintelligence represents the most advanced and powerful stage of AI development, where machines could exceed human intelligence and significantly shape the future of humanity.

9. Asset Compression

The process of reducing the number of megabytes required to store AI-related files such as models, images, or datasets without losing essential information.

10. Available Memory

The amount of free memory, measured in megabytes, that an AI system can still use to load data, run models, or perform computations.

Letter B

1. Bandwidth

The amount of data that can be transferred between systems in a given time. In AI systems, bandwidth determines how quickly megabytes of data can move between storage, memory, and processors.

2. Batch Processing

A method where AI systems process large groups of data all at once instead of one item at a time. Batch processing often involves handling many megabytes of data in a single operation.

3. Binary Encoding

The process of representing information using binary values (0s and 1s). All megabytes of AI data are ultimately stored and transmitted using binary encoding.

4. Bit Depth

The number of bits used to represent a single piece of data, such as a pixel or audio sample. Higher bit depth increases data quality but also increases the number of megabytes required.

5. Bitmap Data

A type of image data where each pixel is stored as a set of bits. Bitmap images can quickly grow in size, consuming large numbers of megabytes in AI vision systems.

6. Buffer Memory

Temporary memory used to hold data while it is being transferred or processed. Buffers store data in chunks measured in megabytes to keep AI systems running smoothly.

7. Bulk Data Storage

Large-scale storage designed to hold massive datasets used for AI training or analysis. Bulk storage capacity is measured in megabytes, gigabytes, or larger units.

8. Byte Stream

A continuous flow of bytes transmitted or processed by a system. AI applications often analyze byte streams that accumulate into megabytes of data.

9. Bytecode

An intermediate form of program code that is easier for machines to execute than human-written code. Bytecode files occupy storage space measured in megabytes.

10. Byzantine Fault Tolerance (BFT)

A system property that allows AI or distributed systems to continue functioning even when some components fail or act unpredictably. Implementing BFT requires additional data handling and storage overhead measured in megabytes.

Letter C

1. Cache Memory

A small, high-speed memory area that stores frequently used data so the processor can access it faster. Cache helps reduce the time needed to move megabytes of data from main memory or storage.

2. Capacity Planning

The process of estimating how much memory and storage an AI system will need. Capacity planning ensures there are enough megabytes available to handle data, models, and future growth.

3. Checkpoint File

A saved snapshot of an AI model during training. Checkpoint files are stored in memory or storage and often consume many megabytes so training can resume without starting over.

4. Chunking

The practice of breaking large datasets into smaller pieces. Chunking helps AI systems process megabytes of data more efficiently by handling them in manageable parts.

5. Cloud Storage

A method of storing data on remote servers accessed through the internet. Cloud storage manages files measured in megabytes and allows AI systems to scale beyond local hardware limits.

6. Compression

The process of reducing the size of data so it uses fewer megabytes. Compression helps AI systems store, transfer, and process data more efficiently.

7. Compute Load

The amount of processing work required to run an AI task. Compute load often increases as the number of megabytes being processed grows.

8. Context Window

The amount of data an AI model can consider at one time. Context windows consume memory measured in megabytes, especially in language and vision models.

9. Corpus

A large collection of text or data used to train AI models. A corpus is usually measured in megabytes or larger units depending on its size.

10. CUDA Memory

Specialized memory used by GPUs to accelerate AI computations. CUDA memory holds model data and intermediate results measured in megabytes during processing.

Letter D

1. Data Allocation

The process of assigning a specific amount of memory or storage for data use. In AI systems, data allocation determines how many megabytes are reserved for datasets, models, or tasks.

2. Data Buffer

A temporary storage area that holds data while it is being transferred or processed. Data buffers manage megabytes of information to keep AI systems running smoothly.

3. Data Compression

A technique used to reduce the size of data so it consumes fewer megabytes. AI systems use data compression to save storage space and speed up data transfer.

4. Data Footprint

The total amount of data an AI system uses, including datasets, models, and temporary files. A data footprint is measured in megabytes, gigabytes, or more.

5. Data Loader

A software component that loads data into memory for AI training or inference. Data loaders handle batches of data measured in megabytes.

6. Data Migration

The process of moving data from one storage system to another. During AI system upgrades, data migration may involve transferring large numbers of megabytes.

7. Data Throughput

The rate at which data is processed or transferred by a system. Higher throughput allows AI systems to move more megabytes efficiently.

8. Dataset Size

The total amount of data in a dataset used for AI training or testing. Dataset size is commonly measured in megabytes or larger units.

9. Deep Learning Model Size

The amount of memory required to store a deep learning model, including its parameters and weights. Model size is often described in megabytes to assess efficiency.

10. Disk Storage

Long-term storage used to save files, datasets, and AI models. Disk storage capacity and usage are measured in megabytes, gigabytes, or terabytes.

Letter E

1. Edge Computing

A computing approach where data is processed closer to where it is generated rather than sent to a central server. Edge computing reduces how many megabytes must be transferred over networks.

2. Embedding Size

The amount of memory required to store vector representations used by AI models. Embedding size is often measured in megabytes and affects model performance and storage needs.

3. Encoded Data

Data that has been converted into a specific format for storage or transmission. Encoded data occupies megabytes depending on the encoding method used.

4. Encoding Scheme

A rule set that defines how data is represented digitally, such as text or images. Different encoding schemes can increase or reduce the number of megabytes required.

5. Encrypted Storage

Storage that protects data by converting it into a secure format. Encryption may slightly increase the number of megabytes used due to added security information.

6. End-to-End Data Pipeline

The full path data takes from collection to processing and storage. Each stage of the pipeline manages data volumes measured in megabytes.

7. Energy Efficiency (AI Systems)

A measure of how much energy an AI system uses to process data. Systems that move and store fewer megabytes are often more energy efficient.

8. Execution Memory

The memory required while a program or AI model is running. Execution memory usage is tracked in megabytes during runtime.

9. External Storage

Storage devices located outside a computer, such as USB drives or external hard drives. External storage capacity is measured in megabytes and larger units.

10. Extracted Features

Important data patterns identified by AI models during processing. Extracted features are stored temporarily or permanently in memory measured in megabytes.

Letter F

1. Feature Map

A structured output produced by AI models, especially in image and signal processing. Feature maps are stored in memory and can consume significant megabytes during computation.

2. Feature Scaling

The process of adjusting data values to a consistent range before training an AI model. Feature scaling affects how efficiently megabytes of data are processed and stored.

3. File Allocation Table (FAT)

A file system structure that tracks where files are stored on a disk. FAT helps manage how megabytes are organized and retrieved from storage.

4. File Compression Format

A standardized way to reduce file size so it uses fewer megabytes. Compression formats help AI systems store and transfer data efficiently.

5. File System Cache

A temporary memory area that stores frequently accessed files. File system caches reduce repeated access to storage and manage megabytes more efficiently.

6. Floating-Point Data

Numerical data that includes decimal values and is commonly used in AI calculations. Floatingpoint data requires more bits per value, increasing memory usage measured in megabytes.

7. Frame Buffer

A region of memory that stores image or video frames before they are displayed or processed. Frame buffers often consume large numbers of megabytes in AI vision systems.

8. Fragmentation

A condition where data is stored in scattered locations instead of one continuous block. Fragmentation can reduce storage efficiency even when megabytes appear available.

9. Free Memory

The amount of unused memory available for programs or AI tasks. Free memory is measured in megabytes and determines how much additional data can be loaded.

10. Full Dataset Load

The act of loading an entire dataset into memory at once. This approach can require large amounts of megabytes and is common in AI training scenarios.

Letter G

1. Garbage Collection

An automatic memory-management process that frees memory no longer in use. Garbage collection helps reclaim megabytes so AI programs can continue running efficiently.

2. Generalization Data

Data used to test how well an AI model performs on new, unseen inputs. Storing and evaluating generalization data requires additional megabytes beyond training data.

3. Gigabyte (GB)

A larger unit of data measurement equal to 1,024 megabytes (or commonly approximated as 1,000 MB). Gigabytes are used when datasets or models grow beyond megabyte scale.

4. Graph Data Structure

A way of organizing data using nodes and connections. Graph-based AI systems store graph data in memory and storage measured in megabytes.

5. GPU Memory

Specialized memory used by graphics processing units to accelerate AI workloads. GPU memory holds model parameters and data batches measured in megabytes.

6. Gradient Data

Numerical values calculated during AI training to update model parameters. Gradient data is stored temporarily in memory and contributes to overall megabyte usage.

7. Granularity

The level of detail at which data is stored or processed. Finer granularity often increases the number of megabytes required.

8. Greedy Algorithm

An algorithm that makes the best immediate choice at each step. Greedy algorithms may reduce memory usage by limiting how many megabytes of data must be stored at once.

9. Ground Truth Data

Correct, labeled data used to train or evaluate AI systems. Ground truth datasets are stored in files measured in megabytes or larger units.

10. Growth Rate (Data)

The speed at which data size increases over time. In AI systems, a high data growth rate means storage needs measured in megabytes can expand quickly.

Letter H

1. Hard Disk Capacity

The total amount of data a hard disk can store. Capacity is measured in megabytes and larger units.

2. Hardware Acceleration

The use of specialized hardware to speed up AI processing. Acceleration often reduces how long megabytes of data stay in memory.

3. Hash Table

A data structure that stores information for fast lookup. Hash tables occupy memory measured in megabytes in large AI systems.

4. Hierarchical Storage

A storage design that uses multiple layers such as RAM, disk, and cloud. Each layer manages data measured in megabytes.

5. High-Dimensional Data

Data with many features or variables. High-dimensional datasets often require large numbers of megabytes.

6. Host Memory

Main system memory used by the CPU. Host memory stores AI data and models measured in megabytes.

7. Hybrid Storage

A combination of different storage types working together. Hybrid storage systems manage megabytes across multiple devices.

8. Hyperparameter Storage

Memory used to store configuration values for AI models. These settings are saved in files measured in megabytes.

9. Hyperscale Data

Extremely large datasets used by major AI systems. Hyperscale data is tracked from megabytes up to much larger units.

10. Heuristic Data

Information used by AI systems to guide decision-making. Heuristic data contributes to the overall data size in megabytes.

Letter I

1. Image Dataset Size

The total amount of storage required for image data. Size is measured in megabytes or more.

2. Inference Memory

The memory used when an AI model is making predictions. Inference memory usage is tracked in megabytes.

3. Input Buffer

A temporary area where incoming data is stored. Input buffers hold megabytes before processing begins.

4. Intermediate Data

Data produced during processing steps. Intermediate data can significantly increase memory usage in megabytes.

5. Index File

A file that helps locate data quickly. Index files use additional megabytes to improve performance.

6. Information Density

The amount of useful data stored within a given size. Higher density means more information per megabyte.

7. Initialization Data

Data loaded when an AI system starts. Initialization data occupies memory measured in megabytes.

8. In-Memory Processing

Processing data directly in RAM instead of storage. This approach uses large amounts of memory measured in megabytes.

9. Input Feature Size

The amount of data required to represent input features. Feature size affects how many megabytes are needed.

10. Instruction Cache

A small memory area that stores frequently used instructions. Instruction caches manage small but critical megabytes.

Letter J

1. Job Queue

A list of tasks waiting to be processed. Job queues may store task data measured in megabytes.

2. Joint Dataset

A dataset created by combining multiple sources. Joint datasets often increase total size in megabytes.

3. JSON Data Size

The amount of storage required for data stored in JSON format. JSON files can grow quickly in megabytes.

4. Jitter Buffer

A temporary storage area that smooths data flow. Jitter buffers store megabytes during real-time processing.

5. Job Scheduler

A system that decides when tasks run. Schedulers manage memory and data measured in megabytes.

6. Just-in-Time Processing

Processing data only when needed. This approach can reduce how many megabytes are stored at once.

7. Joint Memory Allocation

Memory shared across multiple AI tasks. Joint allocation must carefully manage available megabytes.

8. Java Bytecode Size

The storage size of compiled Java programs. Bytecode files occupy megabytes on disk. **9.**

Job Metadata

Information describing AI tasks. Metadata adds additional megabytes to total system storage.

10. Junction Storage Point

A logical link between storage locations. Junctions help organize data measured in megabytes.

Letter K

1. Kernel Memory

Memory reserved for core system operations. Kernel memory usage is tracked in megabytes.

2. Key-Value Store

A data storage method that pairs keys with values. Large key-value stores consume many megabytes.

3. Knowledge Base Size

The total amount of stored knowledge in an AI system. Size is measured in megabytes or larger units.

4. K-Means Dataset Size

The amount of data used in clustering algorithms. Dataset size affects memory usage in megabytes.

5. Keras Model Size

The storage required for AI models built with Keras. Model size is measured in megabytes.

6. Keyframe Data

Important frames selected from video data. Keyframe storage contributes to total megabyte usage.

7. Kernel Cache

A cache used by the operating system kernel. Kernel caches store frequently used data in megabytes.

8. Knowledge Graph Storage

Memory used to store relationships between data points. Knowledge graphs require significant megabytes.

9. Kinetic Data Stream

Continuously changing data from sensors or systems. These streams generate growing megabytes over time.

10. K-Nearest Neighbor Memory Usage

The memory required by nearest-neighbor algorithms. Memory use increases as datasets grow in megabytes.

Letter L

1. **Latency**
   The delay between requesting data and receiving it. Latency affects how quickly megabytes move through AI systems.
2. **Layer Output Size**
   The amount of data produced by a neural network layer. Output size is measured in megabytes during processing.
3. **Load Balancing**
   Distributing work across systems so no single device is overloaded. Balancing helps manage megabytes efficiently.
4. **Local Storage**
   Data stored on a device rather than the cloud. Local storage capacity is measured in megabytes.
5. **Log File Size**
   The amount of storage used by system logs. Logs can accumulate many megabytes over time.
6. **Loss History Data**

Stored values showing how model error changes during training. Loss history consumes megabytes.

7. **Low-Precision Data**

Data stored using fewer bits per value. Low-precision formats reduce megabyte usage.

8. **Labeled Dataset**

Data that includes tags or labels for AI training. Labels increase total dataset size in megabytes.

9. **Linear Model Size**

The storage required for a linear AI model. Size is measured in megabytes.

10. **Load Time**

The time required to bring data or models into memory. Load time depends on megabytes involved.

Letter M

1. **Memory Allocation**

Assigning a specific amount of memory for tasks. Allocation is tracked in megabytes. 2. **Memory Footprint**

The total memory an AI program uses. Footprint is measured in megabytes.

3. **Metadata Size**

The storage used by descriptive information about data. Metadata adds extra megabytes.

4. **Mini-Batch Size**

The amount of data processed at once during training. Batch size affects megabyte usage.

5. **Model Checkpoint Size**

The storage required for saved training states. Checkpoints often consume many megabytes.

6. **Model Compression**

Reducing model size while preserving performance. Compression lowers megabytes required.

7. **Model Parameters**

Values learned by an AI model. Parameter storage contributes to total megabytes.

8. **Memory Bandwidth**

The rate at which memory transfers data. Bandwidth affects how fast megabytes move.

9. **Multimodal Dataset**

Data combining text, images, audio, or video. Multimodal data uses large megabytes.

10. **Mutable Data**

Data that changes during processing. Mutable data can increase temporary megabyte usage.

Letter N

1. **Network Bandwidth**
   The capacity of a network connection. Bandwidth controls how many megabytes can be transferred.
2. **Neural Network Size**
   The total storage required for a neural network. Size is measured in megabytes.
3. **Node Memory**
   Memory available on a computing node. Node memory limits megabytes per task. 4.
Normalization Data
   Values used to scale data consistently. Normalization adds to dataset megabytes.
5. **Non-Volatile Memory**
   Memory that retains data without power. Capacity is measured in megabytes.
6. **Numerical Precision**
   The detail level of stored numbers. Higher precision increases megabytes.
7. **Noise Data**
   Unwanted or random information in datasets. Noise increases data size in megabytes.
8. **Neural Activation Size**
   Memory used to store neuron outputs. Activation size is measured in megabytes.
9. **Network Transfer Size**
   The amount of data sent across a network. Transfer size is tracked in megabytes.
10. **Node Cache**
    Temporary storage on a node for faster access. Cache size uses megabytes.

Letter O

1. **Object Storage**
   A storage method that manages data as objects. Capacity is measured in megabytes.
2. **Offline Dataset**
   Data stored locally rather than streamed. Offline datasets occupy megabytes.
3. **On-Device Storage**
   Storage located on the AI device itself. On-device capacity is measured in megabytes. 4.
Operational Memory
   Memory used during active AI operations. Usage is tracked in megabytes.
5. **Optimization Data**
   Information used to improve model performance. Optimization data uses megabytes.
6. **Output Buffer**
   Temporary storage for results before delivery. Buffers hold megabytes.
7. **Overhead Data**
   Extra data required for system operation. Overhead increases total megabytes.
8. **Object Detection Output Size**

The amount of data produced by detection models. Output size is measured in megabytes.

9. **Online Storage**
Storage accessible via the internet. Online storage limits are measured in megabytes.

10. **Open Dataset Size**
The storage footprint of publicly available datasets. Size is measured in megabytes.

Letter P

1. **Parameter Count**
The total number of model values. Count correlates with megabytes required.

2. **Persistent Storage**
Long-term storage that retains data. Capacity is measured in megabytes.

3. **Pipeline Buffer**
Temporary storage between processing stages. Buffers manage megabytes in motion.

4. **Preprocessing Data**
Data created while preparing datasets. Preprocessing adds megabytes.

5. **Precision Reduction**
Lowering numeric detail to save space. Reduction decreases megabyte usage.

6. **Prediction Output Size**
The amount of data produced by model predictions. Size is measured in megabytes.

7. **Primary Memory**
Main system memory used for processing. Capacity is measured in megabytes.

8. **Processing Overhead**
Extra resources needed during computation. Overhead increases megabyte usage.

9. **Profiling Data**
Information collected to analyze performance. Profiling data uses megabytes.

10. **Paged Memory**
Memory divided into fixed-size blocks. Paging manages megabytes efficiently.

Letter Q

1. **Quantization**
A   technique that reduces numerical precision to lower memory usage. Quantization helps AI models use fewer megabytes.

2. **Query Cache**
Stored results from previous searches or requests. Query caches save megabytes by avoiding repeated computation.

3. **Query Latency**

The time it takes to retrieve data after a request. Latency depends on how many megabytes must be accessed.

4. **Queue Buffer**

   A   temporary holding area for data waiting to be processed. Buffers store megabytes in order.

5. **Quick Access Memory**

   Memory designed for fast retrieval. It handles megabytes needed immediately by AI tasks.

6. **Quantitative Dataset**

   Data made up of numerical values. These datasets are stored and measured in megabytes.

7. **Query Result Size**

   The amount of data returned from a query. Result size is measured in megabytes.

8. **Quorum Storage**

   A   distributed storage method requiring agreement across systems. Data is replicated in megabytes.

9. **Quality Metrics Data**

   Information used to measure model performance. Metrics consume storage in megabytes.

10. **Quantized Model Size**

    The reduced size of an AI model after quantization. Size is typically measured in megabytes.

Letter R

1. **RAM Usage**

   The amount of active memory in use. RAM usage is tracked in megabytes.

2. **Read Throughput**

   The speed at which data is read from storage. Throughput affects how fast megabytes are accessed.

3. **Real-Time Processing**

   Processing data immediately as it arrives. Real-time systems manage megabytes continuously.

4. **Recursive Memory Use**

   Memory consumption during repeated function calls. Usage accumulates in megabytes.

5. **Redundant Storage**

   Extra copies of data kept for reliability. Redundancy increases total megabytes stored.

6. **Resource Allocation**

   Assigning system resources like memory and storage. Allocation is measured in megabytes.

7. **Runtime Memory**

   Memory required while a program is running. Runtime memory usage is tracked in megabytes.

8. **Raw Dataset Size**
   The size of unprocessed data. Raw datasets often require many megabytes.
9. **Read-Only Memory Footprint**
   The storage space used by non-modifiable data. Footprint is measured in megabytes.
10. **Retrieval Speed**
    How quickly stored data can be accessed. Speed depends on the number of megabytes involved.

Letter S

1. **Sample Size**
   The number of data examples used in AI training. Larger samples increase megabytes.
2. **Scalable Storage**
   Storage that grows with demand. Scalability supports increasing megabytes.
3. **Serialized Data Size**
   The storage size of data converted into a transferable format. Size is measured in megabytes.
4. **Shared Memory**
   Memory accessible by multiple processes. Shared memory usage is tracked in megabytes.
5. **Snapshot Size**
   The storage required for saved system states. Snapshots consume megabytes.
6. **Sparse Data Representation**
   A storage method that saves space when data contains many empty values. Sparsity reduces megabytes.
7. **Storage Capacity**
   The total amount of data a system can store. Capacity is measured in megabytes or more.
8. **Streaming Data Buffer**
   Temporary storage for incoming data streams. Buffers handle megabytes in motion. 9.
System Cache
   Memory that stores frequently used data. Caches manage megabytes for faster access.
10. **Synchronization Data**
    Information used to keep systems aligned. Sync data adds to total megabytes.

Letter T

1. **Temporary Storage**
   Short-term storage used during processing. Temporary storage holds megabytes briefly.

2. **Tensor Size**
   The memory required to store multidimensional data structures. Tensor size is measured in megabytes.
3. **Throughput Rate**

The amount of data processed per unit time. Rate is often described in megabytes per second.

4. **Training Dataset Size**

   The total amount of data used to train AI models. Size is measured in megabytes.

5. **Transfer Buffer**

   Temporary memory used during data movement. Buffers hold megabytes during transfer.

6. **Transactional Storage**

   Storage that ensures data consistency. Transactions add overhead measured in megabytes.

7. **Thread Memory Usage**

   Memory consumed by execution threads. Usage is tracked in megabytes.

8. **Token Storage Size**

   The memory required to store processed tokens. Token data uses megabytes.

9. **Time-Series Data Size**

   The storage footprint of data collected over time. Size increases in megabytes.

10. **Training Checkpoint Size**

    The storage required for saved training states. Checkpoints are measured in megabytes.

Letter U

1. **Unified Memory**

   A memory system shared between CPU and GPU. Unified memory manages data in megabytes efficiently across processors.

2. **Upload Size**

   The amount of data sent to storage or the cloud. Upload size is measured in megabytes. 3.
User Data Storage

   Storage allocated for user-generated files. Capacity is tracked in megabytes.

4. **Utilization Rate**

   The percentage of memory or storage currently in use. Utilization reflects how many megabytes are occupied.

5. **Uncompressed Data Size**

   The size of data before compression. Uncompressed data often uses more megabytes. 6.
Update Payload

   Data sent during software or model updates. Payload size is measured in megabytes.

7. **Uptime Storage Logs**

   Logs recorded while systems run continuously. Logs accumulate megabytes over time.

8. **Unified Dataset**

   A dataset combined from multiple sources. Size is measured in megabytes.

9. **User Session Memory**

   Memory allocated for active users. Session memory uses megabytes dynamically.

10. **Usage Quota**

    A storage limit set for users or systems. Quotas are defined in megabytes or larger units.

Letter V

1. **Validation Dataset Size**
   The amount of data used to evaluate AI models. Size is measured in megabytes.
2. **Vector Embedding Size**
   The memory required to store numerical representations of data. Size is measured in megabytes.
3. **Virtual Memory**
   A system that extends RAM using disk storage. Virtual memory is managed in megabytes.
4. **Volatile Memory**
   Memory that loses data when power is off. Capacity is measured in megabytes.
5. **Video Dataset Size**
   The storage required for video data. Video datasets consume many megabytes.
6. **Versioned Storage**
   Storage that keeps multiple file versions. Versioning increases megabyte usage.
7. **Vision Model Size**
   The memory footprint of computer vision models. Size is measured in megabytes.
8. **Vector Database Storage**
   Storage used for similarity search data. Vector databases consume large megabytes.
9. **Virtualized Storage**
   Storage abstracted from physical hardware. Capacity is tracked in megabytes.
10. **Volume Capacity**
    The total size of a storage volume. Capacity is measured in megabytes or more.

Letter W

1. **Weight Parameters**
   Learned values in AI models. Weight storage contributes to megabytes used.
2. **Working Memory**
   Memory actively used during processing. Working memory usage is measured in megabytes.
3. **Write Throughput**
   The speed at which data is written to storage. Throughput affects megabytes per second.
4. **Warm Cache**
   Cached data that is already loaded. Warm caches reduce repeated megabyte transfers.
5. **Workflow Data Size**
   The total data used across an AI workflow. Size is measured in megabytes.
6. **Web Dataset Size**

The storage required for web-based data. Size is tracked in megabytes.

7. **Weight Compression**
   Reducing model weight size to save memory. Compression lowers megabytes used.
8. **Window Size (AI Context)**
   The amount of data processed at once. Window size affects memory megabytes.
9. **Writable Storage**
   Storage that allows data modification. Capacity is measured in megabytes.
10. **Workload Memory Demand**
    The memory required for tasks. Demand is expressed in megabytes.

Letter X

1. **X-Axis Data Size**
   Data stored along one dimension of a dataset. Size contributes to megabytes used.
2. **XML Data Size**
   The storage footprint of XML-formatted data. XML files often use many megabytes.
3. **XOR Encoding**
   A binary operation used in data processing. Encoded data occupies megabytes.
4. **Execution Context Size**
   Memory required during task execution. Size is measured in megabytes.
5. **External Dataset Size**
   The size of data sourced externally. External datasets add megabytes.
6. **Expanded Memory Use**
   Memory growth during processing. Expansion increases megabytes consumed.
7. **Experimental Dataset**
   Data used for testing AI models. Size is measured in megabytes.
8. **Extraction Buffer**
   Temporary storage during data extraction. Buffers store megabytes briefly.
9. **Extended Storage Capacity**
   Additional storage added to a system. Capacity is measured in megabytes.
10. **Explainability Data Size**
    Data used to explain AI decisions. Explainability outputs consume megabytes.

Letter Y

1. **Yield Data Size**
   The amount of useful output data produced. Yield is measured in megabytes.
2. **Yearly Data Growth**
   Annual increase in stored data. Growth is tracked in megabytes.

3. **YAML File Size**
   The storage footprint of YAML configuration files. Size is measured in megabytes.

4. **Y-Axis Feature Size**
   Feature data along a dataset dimension. Size contributes to megabytes.

5. **Young Generation Memory**
   Memory area for newly created data. Usage is measured in megabytes.

6. **Yield Optimization Data**
   Data used to improve output efficiency. Stored in megabytes.

7. **YourAIStudyBuddy Dataset Size**
   The total data used by AI study tools. Size is measured in megabytes.

8. **Yottabyte Reference Scale**
   A conceptual upper scale of data measurement. Used for comparison beyond megabytes.

9. **Yield Metrics Storage**
   Storage used for performance metrics. Metrics consume megabytes.

10. **Yoked Memory Allocation**
    Linked memory regions used together. Allocation is measured in megabytes.

Letter Z

1. **Zero-Copy Memory**
   A method that avoids duplicating data. Zero-copy reduces megabytes in motion.

2. **Zipped Dataset Size**
   The storage size after compression. Zipped files use fewer megabytes.

3. **Zettabyte Scale (Reference)**
   A large-scale data concept. Used to compare growth beyond megabytes.

4. **Zone-Based Storage**
   Storage divided into regions. Zones manage megabytes efficiently.

5. **Z-Score Normalization Data**
   Statistical values used in preprocessing. Data adds megabytes.

6. **Zero Padding**
   Adding placeholder data to inputs. Padding increases memory megabytes.

7. **Zoomed Feature Maps**
   Scaled representations in vision models. Stored in megabytes.

8. **Zonal Cache**
   Cache divided by region. Cache size is measured in megabytes.

9. **Zero-Day Model Update Size**
   The storage required for emergency updates. Size is measured in megabytes.

10. **Zipped Model Archive**
    Compressed AI model files. Archives reduce megabytes required.

Other Books in the IntelliGloss AI Education Series

The **IntelliGloss AI Education Series** is designed to support school districts, educators, and students as artificial intelligence becomes an increasingly important part of modern life. While this book focuses on helping educators understand how to guide students in the age of intelligent systems, additional books in the series explore artificial intelligence from multiple educational perspectives.

Together, these books help build a structured understanding of AI for classrooms, teachers, and school leaders.

Foundations of Artificial Intelligence

Artificial Narrow Intelligence: The Machines That Learn One Thing at a Time
An accessible exploration of the most common form of AI used today. This book explains how narrow AI systems operate, where they appear in everyday technology, and how students can understand their capabilities and limitations.

Understanding Artificial Intelligence: Concepts Every Student Should Know
A foundational guide introducing the core ideas behind artificial intelligence, including algorithms, machine learning, data, and pattern recognition.

Data, Algorithms, and Intelligent Systems

Algorithms: The Invisible Instructions Running the Digital World
A student-friendly exploration of how algorithms shape search engines, recommendation systems, navigation tools, and many everyday technologies.

Data: Fuel for Artificial Neurons — How Information Powers Intelligent Machines An educational guide explaining the critical role data plays in training AI systems and how data quality influences the reliability of machine learning models.

The Evolution of Artificial Intelligence

From ANI to AGI: Understanding the Path of Artificial Intelligence Development A conceptual exploration of how artificial intelligence has evolved and the difference between current narrow AI systems and theoretical future systems.

Superintelligence: Exploring the Future of Machine Intelligence
An examination of long-term discussions surrounding advanced AI systems and the ethical questions that accompany technological progress.

Responsible and Ethical AI

AI Ethics and Safety: Responsible Innovation in the Age of Intelligent Machines
A guide to the ethical considerations surrounding artificial intelligence, including fairness, bias, accountability, and responsible use.

Artificial Intelligence and Society
An exploration of how AI is influencing industries, decision-making, communication, and global technological development.

AI Literacy for Education

The AI-Ready Classroom *(this book)*
A practical guide designed to help school districts, teachers, and educators introduce artificial intelligence literacy responsibly in grades 6–12 classrooms.

Future books in the IntelliGloss AI Education Series will continue expanding resources for educators, students, and schools seeking to build strong foundations in artificial intelligence literacy.

The IntelliGloss Mission

Artificial intelligence is transforming the world at an extraordinary pace. New technologies are reshaping how people communicate, work, learn, and solve problems. Yet while these technologies are advancing rapidly, many students and educators have not been given the opportunity to clearly understand how artificial intelligence works or how it will influence the future.

The mission of the **IntelliGloss AI Education Series** is to make artificial intelligence understandable, responsible, and accessible for learners of all backgrounds.

IntelliGloss was created to help bridge the knowledge gap between emerging technology and modern education. Through clear explanations, structured learning materials, and thoughtful discussion of ethics and responsibility, the IntelliGloss series provides educators and students with the tools needed to understand artificial intelligence in a meaningful way.

Rather than presenting AI as a mysterious or unreachable technology, IntelliGloss presents it as a human-created system built from data, algorithms, and digital information. By learning these foundations, students gain the ability to think critically about technology and its role in society.

The IntelliGloss approach focuses on three core principles:

Clarity

Complex technological ideas should be explained in ways that students and educators can understand without specialized technical training.

Responsibility
Artificial intelligence education must include discussions about ethics, fairness, and the responsible use of technology.

Empowerment
Students who understand AI are better prepared to participate in a world increasingly influenced by intelligent systems.

The IntelliGloss AI Education Series is designed to support teachers, schools, and students as they explore the foundations of artificial intelligence and its growing role in modern life.

By promoting AI literacy, IntelliGloss aims to help build a generation of learners who understand technology, think critically about its impact, and use it responsibly to shape a better future.

Why AI Literacy Matters for Schools Today

Artificial intelligence is no longer a future concept. It is already shaping the way students learn, communicate, search for information, and interact with technology. From recommendation systems and voice assistants to advanced language models and automated decision tools, AI is becoming part of everyday life.

As these technologies become more common, it is essential that students develop a basic understanding of how artificial intelligence works and how it affects society.

AI literacy helps students:

Understand the Technology Around Them
Many digital tools students use today rely on artificial intelligence. Learning the fundamentals of AI helps students better understand the systems influencing information, communication, and decision-making.

Develop Critical Thinking Skills
AI systems are powerful but not perfect. By learning how AI models are trained and how they process data, students become better equipped to question results, recognize limitations, and think critically about technology.

Use Technology Responsibly
Artificial intelligence raises important ethical questions related to privacy, fairness, bias, and responsible use. AI literacy encourages students to use technology thoughtfully and understand the consequences of digital actions.

Prepare for the Future Workforce

Many industries are beginning to integrate artificial intelligence into everyday operations. Students who understand the basics of AI will be better prepared for future careers across a wide range of fields, not just technology.

Encourage Innovation and Creativity
When students understand how intelligent systems work, they are more likely to explore new ideas, create solutions, and contribute to technological innovation.

Schools have always played a key role in helping students understand the tools and systems shaping society. Just as digital literacy became essential during the rise of the internet, **AI literacy is becoming an essential skill for the next generation.**

By introducing students to the foundations of artificial intelligence, educators can help them become informed thinkers, responsible technology users, and future innovators in an increasingly intelligent world.

School Implementation Guide

The **IntelliGloss AI Education Series** is designed to help schools introduce artificial intelligence literacy in a clear, responsible, and structured way. While each school and district may approach technology education differently, the concepts presented in this book can support a variety of classroom and learning environments.

Educators may use this book in several practical ways:

Technology and Computer Science Courses

Teachers can integrate chapters from this book into existing technology or computer science courses to introduce students to the foundational ideas behind artificial intelligence, machine learning, algorithms, and data systems.

Digital Literacy and Responsible Technology Programs

Many schools already teach digital citizenship and responsible technology use. This book can complement those programs by helping students understand how intelligent systems influence information, online content, and decision-making.

Interdisciplinary Learning

Artificial intelligence connects to many academic subjects. Educators may incorporate discussions from this book into subjects such as mathematics, science, social studies, and economics to help students explore how technology affects multiple areas of society.

Classroom Discussions and Projects

Teachers may use selected chapters as reading material to support classroom discussions, reflection assignments, and research projects. Students can explore topics such as AI ethics, technological innovation, and the role of intelligent systems in everyday life.

Independent Student Learning

Students interested in technology, innovation, and the future of intelligent systems may use this book as a foundation for independent study or enrichment learning.

The goal of the IntelliGloss AI Education Series is to support schools in preparing students to understand and navigate the intelligent technologies that are becoming part of modern life.

Artificial intelligence literacy does not require advanced programming or technical expertise. It begins with helping students understand the ideas, systems, and responsibilities behind the technology shaping their future.

A Letter to School Leaders

Dear School Leaders,

Education has always been about preparing students for the world they will inherit. Throughout history, schools have adapted to major shifts in society—from the rise of industrial technology to the digital revolution. Today, we are entering another transformative era shaped by artificial intelligence.

Artificial intelligence is already influencing how information is created, how decisions are supported, and how many industries operate. Students encounter AI through search engines, recommendation systems, voice assistants, and emerging digital tools. Yet many students graduate without a clear understanding of how these systems work or how they influence the world around them.

Schools now have an opportunity to help students develop **AI literacy**—the ability to understand, question, and responsibly engage with intelligent technologies.

The purpose of *The AI-Ready Classroom* is to support educators and schools in introducing these ideas in a clear, thoughtful, and accessible way. The book does not assume advanced technical knowledge. Instead, it focuses on helping students build foundational understanding, critical thinking skills, and awareness of the ethical and societal implications of artificial intelligence.

Artificial intelligence will continue to evolve, but the need for thoughtful, informed learners will remain constant. By helping students understand the technologies shaping their world, schools empower them to become responsible citizens, innovative thinkers, and future leaders.

Thank you for the important work you do each day in guiding and inspiring the next generation.

With appreciation,

Susie Hala
AI Educator and Author
Creator of the IntelliGloss AI Education Series

AI Literacy Framework for Schools

Artificial intelligence is becoming an important part of the modern world. As students grow up surrounded by intelligent technologies, schools have an opportunity to introduce the foundational ideas that help learners understand how these systems work and how they influence society.

The **IntelliGloss AI Literacy Framework** provides a simple structure that educators can use when introducing artificial intelligence concepts in schools. The goal is not to turn every student into a programmer, but to help students develop the knowledge, awareness, and critical thinking skills needed to navigate an increasingly intelligent world.

1. Awareness

Students begin by recognizing where artificial intelligence appears in everyday life. This includes technologies such as recommendation systems, voice assistants, image recognition, automated decision systems, and intelligent search tools.

2. Understanding

Students learn the basic ideas behind artificial intelligence, including algorithms, data, machine learning, and how computers process information. At this stage, the focus is on clear explanations rather than technical complexity.

3. Critical Thinking

Students explore the strengths and limitations of AI systems. They learn that artificial intelligence can be powerful but also imperfect, and that data, design choices, and human decisions influence how these systems behave.

4. Ethics and Responsibility

Students discuss the ethical questions surrounding artificial intelligence, including fairness, bias, privacy, and responsible use. These discussions help students think carefully about how technology should be developed and used in society.

5. Innovation and Future Learning

As students become more comfortable with AI concepts, they can begin exploring how intelligent technologies may shape future careers, industries, and new forms of innovation.

Suggested Classroom Discussion Questions

Artificial intelligence is a topic that encourages thoughtful conversation and critical thinking. The following questions may help educators guide classroom discussions about artificial intelligence, technology, and the future of intelligent systems.

These questions can be used after reading selected chapters or as part of classroom discussions and group activities.

Understanding Artificial Intelligence

What is artificial intelligence, and how is it different from traditional computer programs?

Where do we see examples of artificial intelligence in everyday life?

Why do computers need data in order to learn?

What is the role of algorithms in artificial intelligence systems?

Critical Thinking About Technology

Can artificial intelligence make mistakes? Why or why not?

Why is it important for humans to understand how AI systems work?

How might artificial intelligence influence the way people work in the future?

What skills might become more important as intelligent technologies become more common?

Ethics and Responsibility

What ethical questions might arise when artificial intelligence is used to make decisions?

How can we ensure that AI systems are fair and responsible?

What role should humans play in overseeing artificial intelligence systems?

Should there be rules or guidelines for how AI technologies are used?

Looking Toward the Future

How might artificial intelligence improve areas such as healthcare, transportation, or education?

What are some potential risks of relying too heavily on artificial intelligence?

How can students prepare themselves for a world where AI plays a larger role?

What responsibilities do future generations have when developing and using intelligent technologies?

Educator Discussion Guide

Suggested Responses for Classroom Discussion Questions

The following responses are provided to support educators in guiding classroom discussions about artificial intelligence. These answers are not meant to be memorized by students but rather to help teachers encourage thoughtful conversation and exploration of ideas.

Understanding Artificial Intelligence

1. What is artificial intelligence, and how is it different from traditional computer programs?

Artificial intelligence refers to computer systems that can perform tasks that normally require human intelligence, such as recognizing patterns, understanding language, or making predictions. Traditional computer programs follow fixed instructions written by programmers, while many AI systems learn patterns from data and improve their performance over time.

2. Where do we see examples of artificial intelligence in everyday life?

Examples include voice assistants, recommendation systems used by streaming services, search engine results, spam filters in email, navigation apps, image recognition in smartphones, and online shopping suggestions.

3. Why do computers need data in order to learn?

AI systems learn by analyzing large amounts of data and identifying patterns within that information. The data acts as examples that help the system understand relationships and make predictions or decisions.

4. What is the role of algorithms in artificial intelligence systems?

Algorithms are step-by-step instructions that guide how a computer processes information. In AI systems, algorithms help machines analyze data, detect patterns, and generate outputs or predictions.

Critical Thinking About Technology

5. Can artificial intelligence make mistakes? Why or why not?

Yes. AI systems can make mistakes because they depend on the data they are trained on and the design choices made by humans. If the data is incomplete, biased, or inaccurate, the AI system may produce incorrect results.

6. Why is it important for humans to understand how AI systems work?

Understanding AI helps people evaluate results, recognize potential errors, and make informed decisions about how technology should be used. Without understanding AI, people may rely too heavily on systems they do not fully understand.

7. How might artificial intelligence influence the way people work in the future?

AI may automate some routine tasks, assist professionals in analyzing information, and help improve efficiency in many industries. At the same time, new types of jobs and skills may emerge as technology evolves.

8. What skills might become more important as intelligent technologies become more common?

Skills such as critical thinking, creativity, problem solving, communication, ethical reasoning, and adaptability may become increasingly important alongside technical knowledge.

Ethics and Responsibility

9. What ethical questions might arise when artificial intelligence is used to make decisions?

Questions may include whether AI systems are fair, whether they respect privacy, how bias in data might affect outcomes, and who is responsible when an AI system makes a mistake.

10. How can we ensure that AI systems are fair and responsible?

Developers and organizations can work to ensure fairness by using diverse and high-quality data, testing systems carefully, monitoring outcomes, and including human oversight in important decisions.

11. What role should humans play in overseeing artificial intelligence systems?

Humans should remain responsible for guiding, supervising, and evaluating AI systems. Human oversight helps ensure that technology is used ethically and aligns with societal values.

12. Should there be rules or guidelines for how AI technologies are used?

Many experts believe guidelines and policies are important to help ensure that AI technologies are used responsibly, safely, and in ways that protect individuals and society.

Looking Toward the Future

13. How might artificial intelligence improve areas such as healthcare, transportation, or education?

AI may assist doctors in analyzing medical images, help improve traffic management and transportation safety, and support personalized learning tools that adapt to students' needs.

14. What are some potential risks of relying too heavily on artificial intelligence?

Potential risks include over-reliance on automated decisions, lack of transparency in how systems work, bias in data, and reduced human oversight in important decisions.

15. How can students prepare themselves for a world where AI plays a larger role?

Students can prepare by developing digital literacy, understanding how technology works, practicing critical thinking, and learning how to use technology responsibly and creatively.

16. What responsibilities do future generations have when developing and using intelligent technologies?

Future generations will play an important role in ensuring that AI technologies are designed and used ethically, responsibly, and in ways that benefit society as a whole.

NOTES:

NOTES:

NOTES:

NOTES: